Economic Growth versus the Environment

© *Global Issues series*

General Editor: **Jim Whitman**

This exciting new series encompasses three principal themes: the interaction of human and natural systems; cooperation and conflict; and the enactment of values. The series as a whole places an emphasis on the examination of complex systems and causal relations in political decision-making; problems of knowledge; authority, control and accountability in issues of scale; and the reconciliation of conflicting values and competing claims. Throughout the series the concentration is on an integration of existing disciplines toward the clarification of political possibility as well as impending crises.

Titles include:

Judith A. Cherni
ECONOMIC GROWTH VERSUS THE ENVIRONMENT
The Politics of Wealth, Health and Air Pollution

Graham S. Pearson
THE UNSCOM SAGA
Chemical and Biological Weapons Non-Proliferation

Michael Pugh
REGENERATION OF WAR-TORN SOCIETIES

Global Issues Series
Series Standing Order ISBN 978-1-349-42564-8
(outside North America only)

You can receive future titles in this series as they are published by placing a standing order. Please contact your bookseller or, in case of difficulty, write to us at the address below with your name and address, the title of the series and the ISBN quoted above.

Customer Services Department, Macmillan Distribution Ltd
Houndmills, Basingstoke, Hampshire RG21 6XS, England

Economic Growth versus the Environment

The Politics of Wealth, Health and Air Pollution

Judith A. Cherni
Research Lecturer
Department of Environmental Science and Technology
Imperial College of Science, Technology and Medicine
London

palgrave

Softcover reprint of the hardcover 1st edition 2002 978-0-333-92956-8

First published 2002 by
PALGRAVE
Houndmills, Basingstoke, Hampshire RG21 6XS and
175 Fifth Avenue, New York, N.Y. 10010
Companies and representatives throughout the world

PALGRAVE is the new global academic imprint of
St. Martin's Press LLC Scholarly and Reference Division and
Palgrave Publishers Ltd (formerly Macmillan Press Ltd).

ISBN 978-1-349-42564-8 ISBN 978-1-4039-0713-4 (eBook)
DOI 10.1057/9781403907134

This book is printed on paper suitable for recycling and
made from fully managed and sustained forest sources.

A catalogue record for this book is available from the British Library.

Library of Congress Cataloging-in-Publication Data
Cherni, Judith A., 1955–
Economic growth versus the environment: The politics of wealth, health and air pollution / Judith A. Cherni
p. cm. – (Global issues)
Includes bibliographical references and index.

1. Air – Pollution – Economic aspects. 2. Air – Pollution – Health aspects. I. Title. II. Series.

HC79.A4 .C484 2002
363.739'2–dc21

2001050096

10 9 8 7 6 5 4 3 2 1
11 10 09 08 07 06 05 04 03 02

To Yuti and Adil

Contents

List of Tables

List of Figures

List of Plates

Acronyms and Abbreviations

APCHS	Air Pollution and Child Health Survey
AQCB	Air Quality Control Bureau, Houston
BACT	Best available control technology
CDC	Center for Disease Control (USA)
CO	Carbon monoxide
HAS	Houston Area Survey
HHSD	City of Houston Health and Human Services Department
HRM	Houston Regional Monitoring
HSA	Health Service Area
MSA	Metropolitan Statistical Area
NAAQS	National Ambient Air Quality Standards
NAMS	National Air Monitoring Station
NO_2	Nitrogen dioxide
NOTF	Nuisance Odors Task Force
O_3	Ozone
PM-10	Particulate matter
PMSA	Primary Metropolitan Statistical Area
PPB	Part per billion
PPM	Part per million
PSI	Pollutant Standard Index
SO_2	Sulphur dioxide
TACB	Texas Air Control Board
TES	Texas Environmental Survey
TSP	Total suspended particles
US EPA	United States Environmental Protection Agency
VOC	Volatile organic compound

Preface

This book was written out of a sense of dismay at witnessing one of the most destructive contradictions of our times, that between society and nature. This conflict between elementary human and biophysical (biological, chemical and ecological) conditions is plainly evident in large, post-modern cities. It is sufficient to walk the streets of any thriving large city in the developed world to perceive the vibrancy of past and current achievements. In particular, wealthy cities today are well known for such things as their luxurious buildings, affluent homes and gardens, well-stocked, concrete superstore malls, particular historical and cultural landmarks, and multi-layered busy motorways. The simple act of breathing in this city we all know and love may also trigger something different, the recognition of an undefined unpleasant smell of dust and smog, or, at worst, an uncomfortable choking sensation. These cities stand encapsulated within enormous thick sheaths of brownish air. Such atmospheric conditions can be distinguished today by viewing cities from a distance or from the air.

Air pollution may well be accounted one of the oldest manifestations of the contradiction between nature and society. It became commonplace during the Industrial Revolution (Brimblecombe, 1988). Usual sources of emissions are, for example, burning coal, metal smelting, power stations, cement works, oil refineries, manufacturing plants, and motor vehicles (Clapp, 1994). Rather than considering the quality of the urban air as an already heavily addressed issue of the past – that is, frequently talked about, thoroughly researched, and subjected to policy – it is treated in this book as a real enigma of contemporary politics. Despite improvements in air quality since the 1940s and 1950s, significant issues have failed to be resolved. These are how best to explain the presence of pollution, how to control its increasing levels, and how to reverse the overall persistent trends that have dominated recent decades. This preoccupation arises from both the ecological, as seen above for the thriving city, and human effects of air pollution. Urban pollution has continued to pose health risks for the inhabitants, as did pea-souper fogs in the past. Yet, we have indeed advanced. Today, remarkable urban air pollution in cities may occur not only in winter, as was the case during the big fogs in the 1950s. For example, severe cases of air pollution, and subsequent death, have occurred during the summer seasons of 1988 in the US, and 1991 and 1994 in the UK.

Two related developments have taken place which have, perhaps inadvertently, displaced the issue of air pollution, and the related consequences of ill health, from the sociological, political and economic arena to the individualistic and purely scientific sphere. We now know that after changes made to the air control legislation in the UK and the US affecting domestic and industrial sources of emissions, urban air pollution had become so low in the 1960s that it was wrongly believed that it could no longer perceptibly influence mortality or morbidity in the population. None the less, it was later found that continuing low-level pollution could not be discounted as the cause of much ill health (Brunekreef, 1997). This major displacement of the issue of pollution was possible because transformations were also occurring at another level. A change in the economic atmosphere of cities was taking place. This change reflected rapid urban growth, increasing industrial activity and the globalization of national and international economies taking off during the 1960s and 1970s. In fact, environmental degradation and its impacts on humans had accelerated dramatically in the US and in Europe in particular, so that by the 1980s, to think of air contamination as solely a minor urban problem would have been grotesque. In the natural sciences, studies started to abound that linked air pollution with ill health. The 1990s marked the re-emergence of acute air pollution in large and economically developed cities, generated by the continuous growth of industry, commercial and financial activity, urbanization, and spiralling levels of energy consumption.

It is argued that any concern and dislike for growth-related pollution has been, in reality, diminished and even buried under the objectives of embracing any chance to enhance cities' and residents' material standards of consumption, and improve cities' international competitiveness. Lack of accountability has thus become endemic, most conspicuously among politicians, entrepreneurs, industrialists and the public at large. The repercussion of individual residents' careless attitude towards the environment is none the less different from an institutionalized environmental negligence that has at times systematically failed to protect the urban environment. The situation of evident prosperity and remarkable pollution sharing the same urban space is evidence of overlooked social processes beyond the natural ones. It remains difficult, however, to assert the ways by which social institutions that create wealth and health improvement have also been significant players in the generation of modern contamination and ill health. To deal with these issues, this book refers to theoretical debates on economic growth and environmental degradation, and on co-evolution of technology, science and environment (Norgaard, 1997). It draws on historical information, monitored data and other empirical information, and significantly, on the perceptions of city inhabitants.

Without wishing to minimize the importance of economic development to societies in the general sense, the argument of this book is about some of the political and economic parameters under which degradation occurs. The principal focus is the concrete biophysical consequences of the way the environment is socially affected. Increasingly, these effects have been interpreted either as global or local in reach. The links between the two geographical levels of reference have been less recognized, an omission which is significant for understanding the unreversed presence of air pollution in major cities. On the one hand, a perspective that connects events of localized contamination to global activities has been rarely sought (cf. Vigar, 2000). On the other, that global economic processes take place, in great measure, in localities has often been forgotten. To explain the reality of the contradiction between the workings of nature and those of current society, as manifested in the local aspect of urban air quality, it was essential to consider a geographical and economic convergence. In large cities, the social, political and economic uniqueness of local places becomes as important as the powerful globalization and growth trends that dominate them.

The second part of the book (Chapters 4–8) brings together descriptions, analyses and discussions of growth and pollution in one city with theoretical conceptions formulated in the first part. The book's structure follows directly from this rationale. Chapter 1 offers an introduction to the magnitude of the problem of air pollution and ill health in world cities, discusses the nature of the contradictions raised and explains the significance of the notion of a second contradiction between society and nature. It introduces sociological and philosophical writing about the role of positivist and realist methods of research and the problems encountered by reductionist thinking. It also outlines the evidence for globalization and the presence of constant air contamination found in world cities, stressing the reality of the problem in spite of a measurable reduction in the levels of contamination in comparison with those that hit cities in the 1950s.

Chapter 2 undertakes a systematic review of the literature that has dealt with the subject. It emphasizes the fact that disciplinary fragmentation has dominated explanations. The chapter examines studies in the natural sciences, the sociology of health, and social and medical geography, garnering useful concepts and crucial information from them. It concludes that an interdisciplinary approach is imperative and that a separation of the fields of interest can be crucial, but only for initiating analysis. It argues against mostly unhelpful dualist and reductionist assumptions while stressing the fact that physical and social realities are continually amenable to mutual influence. A number of basic misconceptions are highlighted which reveal the inherent weakness of mainstream thought. It is concluded that a necessary explanation must also consider the structures of society. The next chapter discusses this dimension.

Chapter 3 outlines the main politics of environmentalist theories that have criticized economic growth for its resultant ecological degradation. It suggests a classification of these theories according to those factors that are claimed to have caused sustained environmental degradation. To develop an appropriate explanation of air pollution and ill health in large, 'world/globe cities', it draws on ecological economics and materialist ecological thought, and on a critical interpretation of globalization. The chapter presents a fresh political and economic approach to contemporary physical environmental conditions and develops an interdisciplinary methodological strategy to address the issue of air pollution in cities.

The theoretical problem of persistent urban pollution and economic growth in the real geographical and historical contexts of a case study is set out in Chapter 4. It discusses changes associated with economic development in the Houston region, the globalization of the oil industry, the making of a world city, and its transformation into the world energy capital. These components have brought considerable wealth to the area while also creating environmental degradation. The chapter raises the subject of the contradictions between economic and urban affluence and the deterioration of nature, and between global competitiveness and local poverty. It compares existing urban landscapes of progress, abundance and glamour with the look and smells of current pollution and urban dereliction. Houston is shown as an epitome of contrasts where wealth and misery stand side by side. Finally, it describes how present local environmental problems, historical developments and the current impact of world economic forces interact with regional growth.

Chapter 5 turns to the issue of air pollution in recent times, examining in greater detail how a place like Houston ranks today among the most polluted cities in the US. It focuses on pollutants, measures of monitored contaminants and the failure to attain the national health safety limits. It also assesses the reports of local contamination by residents in two geographical areas of the city. This chapter indicates the severity of the problem of air pollution in a world city from the two complementary perspectives of lay opinion and scientific expertise. Together with patterns of spatialized risk, it uncovers notably high levels of reported ill health among children and reveals that the extent and type of ill-health symptoms follow a geographical pattern. Further analysis of the book's original database complements the arguments of the next chapters.

Moving on, Chapter 6 looks into whether the remarkable clusters of ill health found in Houston were in any way affected by US health provision. It presents a brief introduction to the US medical care delivery system and the social exclusion that it implies. The chapter exposes the relatively poor achievement levels of public health and the causes of mortality in one of the wealthiest US cities, presenting a detailed socioeconomic analysis of residents' reports of ill health among children. It concludes with a discussion of the importance of conditions of poverty as a trigger for ill health and stresses this well-known aspect of the unequal system of societal organization. This factor alone, however, cannot explain the high incidence of ill health also found in high-income homes. This issue is further analysed through spatial indicators of environmental pollution.

In Chapter 7 the analysis is taken further, into the regulatory fields of industrial emissions. It highlights a number of important flaws in a set of regulations that have promoted a partial betrayal of the declared aims to protect the environment and health. It presents solid evidence for the argument of a persistent trend of pollution due to institutional weakness which, importantly, is rooted in both the economic structure and our configuration of knowledge. Local interests and international producers have forcefully defended restricted government control. The modifying role of household spatial location in relation to sources of industrial air pollution is assessed throughout the chapter while the risk of industrial accidents and severe environmental, and hence health, consequences is highlighted. It underscores the social character of a geographical association between exposure to pollutants and increased risks and stresses the importance of more appropriate regulation to control industrial emissions.

Chapter 8 brings together the themes and issues developed in the book. It discusses several of the lessons that can be learnt from the process of analysing the topic from a political-economy perspective, using an interdisciplinary approach and also a participatory strategy. It identifies the responsibility of social institutions against that of individuals alone to protect the environment and public health. Local and global risk phenomena, and the need to overcome wilful amnesia about previous developments is underlined. The chapter focuses on recent agreements between the US government and the US EPA, and powerful oil multinationals to reduce emissions as a start towards benefiting the local residents, and eventually, the global environment.

Acknowledgements

My greatest debt of thanks is to Dr Andy Pratt of the London School of Economics for his help in focusing many of the ideas presented in this study and for his thought-provoking insights. I would also like to thank Dr Yvonne Rydin of that institution for her useful comments and practical suggestions, Dr Simon Duncan, at Bradford University, who gave me invaluable guidance at the beginning of this project, and Dr Sylvia Chant, also at the London School of Economics, who helped develop my household research. Thanks also to Dr Cathy McIlwaine at Queen Mary and Westfied College London for her valuable insights and constructive comments on three of my chapters. To Jim Whitman, my Palgrave Global Issues Series editor, I am most grateful for a careful reading of my manuscript and for his incisive comments. His personal enthusiasm for the subject and for the book project has been contagious and has on many occasions helped to raise my confidence when it flagged. Further assistance was provided by the expertise of Alicia Felberbaum, who worked with dedication into the night to produce the illustrations, translating my own photographs into their present coherent format. I am grateful too to Lucy Isenberg, who saw the very early draft, and whose firm conviction that this material ought to be published greatly encouraged me.

The successful completion of this research would not have been possible without the cooperation and assistance of many people in Houston. I am particularly grateful to those many anonymous residents who so willingly responded to my enquiries and whose comments served to endorse further suspected environmental and human risks from globalization and to shape a conceptual view for this kind of participatory knowledge. In this respect I would also like to mention the assistance of Gene McMullen at the City of Houston Bureau of Air Quality; S. Hardikar at the City of Houston Health and Human Services Department, and Professor Stephen Klineberg of Rice University, Houston, who kindly gave of their time to my interviews. A number of very resourceful libraries were consulted in Houston. A most useful institution was the

M.D. Anderson Cancer Center Library at the University of Texas. No less important were the libraries in the School of Public Health, also at the University of Texas; the well equipped library of Rice University, the University of Houston Library and the main Public Library in Houston. In the UK, the main libraries to assist in my researches were the British Library of Political and Economic Sciences at the London School of Economics, the Science Reference and Information Service of the British Library, Local Libraries, and the Inter-library Loan Service for much of the medical material.

Finally, and deserving of special mention, are my husband and daughter who helped and supported me from the earliest phases of this investigation until its final production. My husband accompanied me to many interviews in the most dangerous areas of Houston and my daughter's admirable cheerfulness always added colour and warmth to my inevitable periods of incarceration for frenzied writing. I thank them both.

1
Introduction

> The most alarming of all man's assaults upon the environment is the contamination of air, earth, rivers, and sea with dangerous and even lethal materials. This pollution is for the most part irrecoverable; the chain of evil it initiates not only in the world that must support life but in living tissues is for the most part irreversible.
>
> (Carson, 1962, p. 6).

A shocking feature of modern and post-modern cities is that visible economic growth and material prosperity take place irrespective of a simultaneous contamination of the urban environment. The situation appears the more inexplicable if one considers that the well-chronicled human tragedies of the 1950s left no doubt about the potential of urban air pollution to cause illness, even death. As a result of these severe episodes of pollution, stricter control strategies were implemented from then on. The quality of the air was widely considered to have improved, based on the monitoring systems in place and regulatory measures that were generally regarded as successful in limiting pollution's effects on human health. However, despite visible reductions in sulphuric contamination and particles, air pollution has remained for the last half-century a familiar feature of contemporary major cities. At the core of this book is the belief that much of current urban environmental degradation has been underpinned by a long-lasting trend of uncontrolled economic affluence for some, which results in urban pollution and ill health for many more. The book argues that this situation points to a contradiction between the institution of economic growth in the developed Western world and the natural environment that sustains it. Ill health represents a human parameter of such contradiction.

Health and the urban environment

The study of health and the environment has been based on competent scientific research spanning approximately the last twenty years and creating by now a solid body of epidemiological literature. It firmly connects man-made pollutants to changes in human functions (see, for example, Dockery, et al., 1993; Katsouyani et al., 1993; Kinney et al., 1996; Lueunberg et al., 2000; Romieu et al., 1990). Current knowledge on health and pollution, originating in numerous investigations in the natural sciences, is reviewed in Chapter 2. Air pollution has been associated with the impairment of health, particularly in children, the old and the vulnerable. It has also been recognized as a main cause of deterioration of building material and lack of cities' sunlight. Urban living has been singled out as a main cause of the high prevalence of asthma worldwide, particularly among children. Paradoxically, the US, a nation with very high personal income, stands as one of the countries with the highest incidence (Aligne et al., 2000). And, importantly, further epidemiological data show unequivocally that asthma is exacerbated during episodes of air pollution (for example, Ghazi, 1992; Lean, 1994a, 1995; Schoon, 1994a, 1994b). Other studies show an association between increased daily mortality and particle air pollution, for example in Philadelphia, St Louis, Stebenhuille in Ohio and in São Paulo, Brazil (Dockery et al., 1993; Samet et al., 1995; Schwartz and Marcus, 1990; Sunnucks and Osorio, 1992). In the last two cities, the highest mortality rates correlated with the highest annual concentrations of industrial and other sources of air pollution. Further, one of the highest rates of cancer mortality in the US was registered in the 1970s for Baltimore, Maryland, a highly industrialized city (Radford, 1976). Furthermore, pollutants such as sulphur dioxide, airborne allergens, and atmospheric ozone are found in major cities like Athens, Los Angeles, Mexico City, Paris, London and Tokyo. Their presence correlates with symptoms exemplified by, for example, increased emergency patient visits to hospitals, a rise in asthma attacks, more frequent episodes of cardiovascular disease and general respiratory symptoms and, finally, by excess mortality from respiratory problems.

The link between exposure to air pollution and ensuing ill health was not discovered only in the last decades of frenetic urbanization and economic growth. In fact, the phenomenon had already been acknowledged, and also briefly investigated, during the nineteenth century, particularly as a reaction to industrial processes (see International Labor Office, 1927; Kapp, 1950; Spence et al., 1954). It was only in the last decades of the twentieth century, however, that scientists and policy-makers focused more tenaciously on the consequences of escalating disarrangement in ecological and human functions due to industrial and traffic toxic emissions (Chapter 2). Despite a considerable body of epidemiological studies, regulatory change and technological advance, air pollution in cities has persisted as an urban problem. Its simple presence is pointing towards a grotesque misjudgment. In different ways, and with varying degrees of responsibility, most politicians, scientists, industrialists, and the members of the public alike have avoided looking at current urban pollution as a complicated biophysical condition with social roots. Scientific evidence attesting chronic and also fatal health consequences of air pollution in past and recent years has only encouraged a limited success in rectifying the attribution of the problem to include social roots. The role of the lay population in this oversight has been important. Air pollution has been interpreted as endemic and unavoidable, a part of progress, and a reasonable externality that society must endure in exchange for the benefits of economic growth. For many, however, accepting air pollution today in their own residential areas has become a less attractive proposition (see Chapter 5).

There have been great smog disasters in the past, which testify to the extremely adverse consequences of air pollution. Memorable episodes took place in the Meuse Valley, Belgium in 1930; and in Donora, Pennsylvania, in the US in 1948. In the UK, episodes occurred during the London Big Smoke, in 1952; in 1873, it was estimated that there were 700 more deaths than normally expected in London at that time of year, and in 1892, an estimated 1000 more deaths (Brimblecombe, 1988). Disregard for the natural environment has been constant over time. It was not until the above-mentioned extreme air pollution episodes struck large cities in the 1950s that the UK and US governments considered the evidence of such devastating health effects of air pollution for regulating toxic emissions. In the decades after the Big Smoke, incidences of extremely hazardous air pollution have decreased considerably in American and European cities. However, sources of toxic emission have multiplied exponentially and, in numerous cities, the levels of a different type of contaminant have risen dramatically. In the 1970s, American and European legislators considerably tightened controls on air pollution from the so-called 'stationary' sources of homes, commerce and industry. Unquestionably, this has led to improvements in many aspects of urban air quality. However, such improvements, arising as they do from the overall decline in domestic and industrial emissions, have been offset to a large extent by remarkable industrialization, emissions from widespread usage of motor vehicles, and the expansion of cities (Middleton, 1999; Parliamentary Office of Science and Technology, 1994; UNEP and WHO, 1992).

By the 1960s and 1970s, after changes made to the air control legislation, it was mistakenly thought that ambient concentrations of air pollution no longer had any discernible influence on health in the population. None the less, a series of studies has shown that even at low concentrations, pollutants were still associated with day-to-day changes in concentrations of air pollution that affected health. Indeed, a strong correlation between the number of deaths and the presence of high levels of total suspended particles (TSP) and sulphur dioxide (Schwartz and Marcus, 1990) was uncovered during the London winters of 1958–72. Significantly, the strength of the correlation remains despite the fact that absolute levels of mortality and air pollution were lower than in the winters of earlier years. In cities with high levels of air pollution, mortality rates seem to be higher than in less polluted cities (Lawrence, 1993; US EPA, 1990a).

Both successes and failures to reduce air pollution have occurred. Yet on balance the increasing trends have not yet been reversed. Today, the high incidence of respiratory and other diseases originating from exposure to levels lower than extreme, in addition to the reasons for fewer excess deaths, calls for a thorough approach to the association between current pollution and ill health. The reappearance in recent years of pollution episodes in big cities potently indicates that, over almost half a century, political and economic agendas have displaced the need and responsibility to protect the urban environment, and, hence, human health.

The atmospheric conditions in London in December 1991 closely resembled those of the winter days of the London Fog; and in 1994, polluted air almost choked Britain. During the first episode, pollution from car exhausts, combined with freezing temperatures produced smog that exceeded the levels of many of the world's dirtiest cities (Palmer and Ballantyne, 1991). In those days, the government urged motorists to stop driving as winter air pollution reached the highest levels since records began in 1976 (Brown, 1991). The UK Department of the Environment announced that the health risks from high levels of nitrogen and sulphur dioxide would make it dangerous for asthmatics, bronchitics, the elderly and babies to go outdoors until the following day (Lonsdale and Lockhart, 1991). Following the other pollution episode in July 1994, the biggest asthma outbreak ever registered anywhere in the world until then occurred over much of England (Lean, 1994c). An estimated 160 excess deaths were registered, while hundreds more people experienced breathing difficulties (Connor, 1994; Lean, 1994b). The scale of the outbreak almost certainly meant pollution was a factor (Hall, 1994), since this happened as Britain was enduring the dirtiest air summer recorded in that decade (Lean, 1994a). Unusual climate conditions were discounted as a possible cause of such episodes because there had been plenty of thunderstorms in the past but none had led to such episodes (Lean, 1994b; National Asthma Campaign, 1994). It was the high pollution levels in the previous weeks that had sensitized people's lungs (British Lung Foundation, 1994, in Lean, 1994c). The recurrence in the 1990s of remarkably high levels of air pollution in one of today's leading financial and business global cities, London, strongly suggests that historical memory of past episodes is now a political necessity in order to reverse current urban air pollution.

It is apparent that neither environmental legislation nor higher levels of personal and national income in the US and advanced European countries have guaranteed a cleaner environment for all citizens. Yet few studies are available on the links between environmental degradation and ill health on the one hand, and on the other, between the political and economic structures, which, over time, have enabled air pollution to remain a perpetual feature of contemporary cities. In summary, the study of health and the environment has an extensive and venerable career involving a number of disciplines, among which the social and political sciences have been the most notable for their absence. Hence, rethinking urban pollution in world cities and relating it to the emergence of regional industrialization, weak regulatory instruments and the economic direction of uncontrolled growth becomes a prerequisite for understanding the current presence of local air pollution and the failure to reverse the existing trend.

Growth, globalization and world cities

Since Carson's (1962) resonating condemnations, there has been a series of influential publications that have highlighted tension between economic growth and the environment (see, for example, Bartelmus, 1986; Jacobs, 1994; Meadows et al., 1972; Pearce et al., 1989; Pearce, 1991; Watts, 1983; WCED, 1987). While one strand within this literature has focused on resource depletion (Blaikie, 1985; Hecht and Cockburn, 1989; Prestt, 1970), the other has concentrated on the contamination of nature (Bertell, 1985; Blowers, 1989; Blowers and Lowry, 1987; EMEP, 1997; ENDS, 1994b; Ives, 1985; World Resources Institute, 1994). This book belongs into the second strand and it aims to develop a social and political explanation of the persistent problem of air pollution in major cities, and its main repercussions on health. It applies a political-economy perspective that connects transformations in the natural environment and in health to the institution of economic growth and the latest neo-liberal process of globalization.

The connection between economic growth and environmental degradation has been an issue for critical debate since the 1960s. Influential theories, from the limits to growth approach to the latest developments in ecological modernization, are discussed in the first part of Chapter 3. Illuminating insights for an understanding of the complexity of the topic are found in writings by authors in radical sociology and ecological economics. Foremost, we draw on Herman Daly's advocacy for a steady-state economics as an antidote to "growthmania" (Redclift, 1996). Daly (1992) writes:

> The economy grows in biophysical scale, but the ecosystem does not. Therefore, as the economy grows, it becomes larger in

> relation to the ecosystem. Standard economics does not ask how large the economy should be relative to the ecosystem.
>
> (p. 180)

Our composite view can be illustrated by noting that 'sustainability as a desirable objective has served to obscure the contradictions that development implies for the environment' (Redclift, 1987, p. 2). The co-evolutionary theory proposed by Richard Norgaard offers a useful framework for the links between biophysical and social transformations. Co-evolution in biology assumes ecosystemic relations between the larger process of modernization, specifically, technological change and globalization, and the predicted, but also unforeseen, environmental consequences (Norgaard, 1985, 1997). This book takes into consideration the process of globalization because, in the last decades, it underpins many of the local biophysical alterations in major cities and governments' political postures. Keeping clean air and healthy environments has been particularly difficult under conditions of economic growth and globalization. These insights are discussed in the second part of Chapter 3.

The way we live today, the paths used to relate to nature, the type of commodities produced, the features of international trade, and, importantly, the quality of the air we breathe, have gone through some significant modifications. Rather than being the result of isolated events of either local or nationally promoted economic policy, these changes portray a link between the previously existing institution of economic growth and the global reorganization of the national and international economy. As part of globalization, reorganization by firms took place during the 1960s and 1970s, and acquired strength in the 1980s and 1990s, and has been accompanied by legislative support of neo-liberal policies such as the deregulation of national markets. For a definition, according to the Organization for Economic Co-operation and Development (OECD, 1997), globalization can be thought of as a process in which economic markets, technologies and communication patterns gradually exhibit more global characteristics and fewer 'national' or 'local' ones (p. 19). This elucidation seems sufficient, but only for the task of describing how globalization appears, rather than why it takes place. Globalization has arisen as a powerful, and abrupt, strategy to improve growth-related profit. It has originated in a series of economic and political crises marking the end of the post-war boom of the 1960s and following the oil crisis of the 1970s. The increasing internationalization of finance and the economic relations of production and the deregulation of trade best illustrate the process of globalization. As capital markets become global, the fate of whole countries' economies can fall prey to investors in the international money markets (Yearley, 1996).

The importance of globalization becomes even more evident when we look at the global city literature. Globalization emerges as the common explanatory factor in the accounts of the appearance of the contemporary global or world city phenomenon (King, 1990). Soja presents a convincing theoretical account of the world city drawing on Mandel's (1978) arguments on long wave restructuring of the economy (Soja, 1986; Soja et al., 1983; see also Cohen, 1981; Sassen, 1991). At different historical periods, it is argued, major and minor restructuring of capital has taken place, which has had different effects at different levels of the world economy, the national, the sub-national, and the city. In the last decades, however, a much more dramatic internationalization of capital has taken place, leading to a more widely distributed system of production. In this context of unprecedented social transformations, the global or world capitalist city has emerged as the centre for the financial management, international trade, and corporate headquarters of radiated tentacles of international corporations. 'Major cities, as the places where this politico-economic specialization is grounded physically, are the cotter pins holding the capitalistic world-economy system together' (Feagin, 1985, p. 1210).

World cities occupy varied functions in the globalized economy of the end of the twentieth and beginning of the twenty-first centuries. Feagin (1987) distinctly argues that cities are not discrete and independent entities, but rather interconnected parts of a globally integrated capitalist economy, with many cities fulfilling different roles. In this sense, large cities are not islands unto themselves; rather they are situated places greatly affected by capital investment flows within the regional, national and international contexts (Feagin, 1988). Such are the California Silicon Valley technology centre and the Houston Ship Channel petrochemical complex, both in the US. The contradiction between economic growth and environmental degradation is analysed for Houston, a city that provides raw materials, specialized services and markets for many other economies in the world, and this role is essential in the internationalization of the economy. This city is home both to one of the world's largest and most comprehensive oil complexes and to an internationally renowned medical centre (Chapter 4). Typical of this world city at the end of the twentieth century are intensive petrochemical manufacturing, financial activity and advanced technology industries.

The book shows that, at the same time as achieving growth, global reach and the creation of enormous wealth, Houston has attained one of the highest US levels of air contamination (Chapter 5). Moreover, unequal access to health services and poor public health achievements stand in stark contradiction to the outstanding quality and huge size of the 'medical industry' in Houston (Chapter 6). A major failure that needs to be addressed, however, has been a general reluctance to question the ways in which places and world cities have become more or less polluted by the combined actions of ecological laws and political and economic forces in society. It is precisely this latter inadequacy which I wish to address in the following chapters, which discuss the contrast in world cities between a growth paradise and a hell of pollution. The last part of Chapter 3 develops the conceptual and methodological tools to undertake the task.

The role of the sciences and critical realism

There exists a wide range of useful theoretical formulations and empirical analyses on the ecology, chemistry and biology of health and environment. There is also a vast literature on micro-sociology of health factors that are associated with the chances of contracting illness and on the parameters of disease distribution and on the importance of place (Chapter 2). A crucial point is that the study that comprises the environmental health agenda has presently come to be dominated by underlying geographical perspectives which are regionally constituted and perceived as an analysis of location-specific deployment patterns (Jones and Moon, 1987). These have usually been ascribed to the fields of epidemiology and medical geography. A problem that has emerged with this literature is that there is little common ground among such perspectives, the intersections are few, and no single approach addresses the enquiry in its totality. When the issue of urban air pollution has been approached as a predominantly biophysical subject, with the objective either of explaining or solving it, it has proved very complex for the natural sciences. Equally, the issue has proved difficult for the social sciences to evaluate and fix, due to the commitments underlying our political institutions, our entrenched alienation from nature, private-versus-public confrontations, and the powerful status of the market in society. An interdisciplinary and participatory perspective on the subject is necessary. There are important lessons to be learned from the natural and social sciences, from an interpretation of the political priorities and economic forces at play, and from the say of the public. Comprehensive and politically informed knowledge of this type will be important to improve situations where biophysical and social phenomena converge.

Reviewing the current state of the art of theoretical modes of analysis of contemporary urban pollution and ill health shows that the essential information revealed in the sciences originates in positivist methodologies. These consist of tools to identify, measure and control the biophysical transformations in nature and in human beings. Even though we fully acknowledge that the information obtained by these methods is critically important, the methodological positivism of the relevant sciences proves very narrow for our explanations. Positivist approaches do not look beyond the descriptive and mechanical aspects of the relationship between air pollution and ill health. On the sociology of health front, much of their research is carried out on a reductionist basis. It has focused almost exclusively on health changes in relation to the effects of either household socioeconomic circumstances, care institutions, or the characteristics of the living area. However, one should not underestimate the value of these studies. Yet, because they are very focused, their explanations constitute only a part of a complete analysis.

The role of the sciences in explaining the linkages between both economic growth and environmental degradation has come under increasing scrutiny from many authors due to the sciences' inadequacy for the task (for example, Beck, 1997; Caldwell, 1977; Oakley, 1992; Redclift, 1984; Yearley, 1991a, 1991b). Principal philosophical difficulties for producing competent knowledge, that is, knowledge that can be politically implemented for social benefit, are that the foundations of the separation between the social and the natural sciences, based as these are on dualistic or separationist modes of thought, go very deep (Dickens, 1992; see below). Dualisms are in a very important respect organizing categories, both shaping scientific thought and research, and structuring everyday non-scientific and commonsense contexts of thought. We have thus a dichotomous understanding: one based on social theory, the other on natural science (Benton, 1994). Epistemological limitations of this type can restrict the freedom of analysis, particularly when approaching our subject with well-defined biophysical and also social dimensions.

Additionally, this book points to the extent that research on pollution, health and economy indicates that most available research models have often emphasized and reproduced reductionist and also ahistorical views of society. These have minimized the social origins of environmentally related health problems that afflict contemporary society. In this manner, widely held beliefs about the short- and long-term effects of hazardous environments are necessarily incomplete and the political alternatives for dealing with them have turned out to be not as effective as might have been expected. It is argued that a 'disintegrated' view of the problem, and a non-participatory approach that excludes people from the analysis, will be doomed to produce results similar to those of previous work and hence will delay implementation of measures to bring permanent improvements in the quality of urban air.

Instead, the scientific method of critical realism as developed particularly by Sayer (1992) is employed. As opposed to positivism, in critical realism the study of social reality implies the examination of both observable and non-observable events, that is, the underlying processes of causality. The book operates on the principle that to establish knowledge we also need to recognize the power of things to cause events and give rise to ways of acting. Critical realism recognizes the domains of mechanisms and processes of events and experiences. Following the principles of critical realism, methods for quantification and qualification to assess the events are employed. In practice, critical realism does not offer one particular model of research. Pratt (1995) emphasizes that

> The most exciting moment in the development of critical realism should be the attempt to practically work through its implications in order to understand the world. ... A key sticking point in the practical application of critical realism is research methodology. (p. 67)

The book consists of theories, political analysis of historical and current economic development, scientific data, the participatory database of a case study and the views of key professional actors. One point of departure is the concern that many researchers have expressed with traditional areas of environmental health, such as distribution of air pollution, and the role of household conditions in triggering disease (Chapter 2). A further departure point is the discussions surrounding the issue of local economic growth and environmental problems from the less researched perspective of global reorganization of the economy (Chapter 3). Within our limitations, we have attempted to transcend the restrictions of ahistoricism inherent in current empirical fieldwork. On the other hand, abstract theoretical notions alone may cause difficulties for actual political implementation. Therefore, we have expanded structural theoretical concerns with capitalism by incorporating the concrete and conflictive relations between society and nature (Chapter 4). The venue of study is the international city of Houston and four socioeconomic and geographic clusters of residents within the city (Chapters 5, 6 and 7).

Contradictions of economic growth

The notion of a contradiction between nature and society has been mentioned in Marx and Engels's writings, but only later writers developed a more satisfactory analysis. The so-called 'second contradiction of capital' implies that, in addition to the traditional focus on a first contradiction (that is, the mechanisms of exploitation of labour power within growing industrial economies), there are also 'social mechanisms of destruction and exploitation of nature' (O'Connor, M., 1994, p. 7). By arguing that 'capitalist agriculture is harmful to the soil', Marx already assumed that capital undervalues nature (O'Connor, J., 1998). However, 'he failed to argue that "natural barriers" may be capitalistically produced barriers, that is, a "second" capitalized nature' (O'Connor, J., 1998, p. 160). Capitalization of the conditions of production in general, and the environment and nature in particular, tends to raise the cost of capital and reduce its flexibility. For example, 'the use of pesticides in agriculture at first lowers, then ultimately increases, costs as pests become more chemical resistant and also as the chemicals poison the soil' (O'Connor, M., 1994, p. 6). Foster (1992) sees capitalism as digging its own grave, tending to maximize the overall toxicity of production and promoting accelerated habitat destruction. It is argued that the phenomenon of air pollution, and consequent environmental ill health, points to another concrete manifestation of this notion of the second contradiction of capitalism. Yet, neither theories of air pollution nor of health had been used to foster its understanding. What is even less understood is how two definitively different mechanisms involved in the second contradiction, one biophysical, the other social, keep in continuous interaction through political and economic structures.

As a basic theoretical point of departure, the concept of a second contradiction of capitalism is insufficient on its own, however, and there is need for further clarification. Dickens (1997) rightly questions whether it is correct to give such weight to capitalism in the destruction of the environment. Discounting a yes for an answer, a further philosophical component must be taken into consideration. Drawing on Marx's earlier work, he develops an understanding of our alienated relations with nature, which partly explains why we appear not to care about environmental degradation. According to Dickens (1997), the division of labour in modern society is responsible for a great deal. It divides an understanding of society–nature relations into the fragmented understandings offered by different and well-established disciplines. Just as importantly, it 'divides people up in to "lay" and "expert", ensuring that their two types of knowledge remain segregated from one another' (Dickens, 1997, p. 191). Indeed, drawing on social theory and scientific knowledge is not enough to explain the condition of enduring urban pollution in so-called world cities. This book attempts to amalgamate 'lay' knowledge from the public with essential scientific information. Therefore, it draws on an empirical case study from which crucial lessons were learned.

A strong sense of bewilderment has impelled this book because contamination in the cities of developed countries has become ever more difficult to define and explain in other than traditional natural science terms. Other social environmental problems with strong biophysical features, like, say, food scarcity, land erosion and poverty in countries of the Third World, or for that matter, the loss of biodiversity and international trade, have occupied a clear and defined niche in the political-economy agenda of environmentalism. But urban pollution and health risk, the latter defined principally as a medical condition, have mainly been omitted from the same agenda. Further, practical and technical difficulties exist because the concentrations of pollutants that have been declared as safe have, occasionally, turned out to be erratic. The facts that there are now so many interacting pollutants, and that the level of concentration of air contaminants is highly affected by climatic and topographic variables, make the isolation of environmental factors the most difficult to measure and reproduce in relation to health. Human beings are usually exposed to many factors, such as poverty, deficient health care provision, and individual and group unhealthy lifestyle which may also trigger ill health.

To a large extent, my position draws on the work of experts in the social sciences. These have already identified a mounting environmental crisis, looked into many of its consequences, criticized our social ways of production and running our lives and, not least, have highlighted the inadequacies of theoretical models to interpret them. However, in a wide sense, practical concerns over pollution and health matters had made their appearance most commonly in disciplines such as medicine, epidemiology, toxicology, chemistry and ecology. Conceptualization of the biophysical phenomena in social, political and historical terms has received a lower profile than other problems. In addition, due to a preconceived dualist epistemology, as mentioned above, the undeniable ecological and medical character that distinguishes pollution and health has apparently promoted its own unnecessary exclusion from mainstream political debates within environmentalism.

Natural and social scientists are culprits for these exclusions and omissions. Natural scientists have generally reduced the problem to its biological dimensions, overlooking the powerful influence of social institutions. Social health scientists rely on indicators constructed from poverty-related conditions, including housing location, to determine the chances of individuals or groups suffering ill health. The focus of the sciences on exclusive aspects of the problems has contributed to a tacit, and nonsensical, underestimation of the vital linkages that connect biophysical events and social structures that are found beyond the boundaries of the individual or the household. It is necessary therefore to build knowledge on the basis of deeper and structural connections so as to produce information that is practical for political decision-making. This is necessary because, for policy-makers, the social structures that have promoted regional transformations and globalization seem to be unproblematic and unrelated to the quality of the urban air, and, even less to public health. For this reason, technical, rather than political, considerations such as emissions control, often override attempts at explaining the problem from a more complex perspective. A main objective of this book is to trace these vital, yet difficult to visualize, links of causation.

There is no doubt that urban environmental sustainability represents a beneficial and desired aspiration. The contention of this book is that unless the social reasons for urban degradation are properly identified (through combined participatory, political and economic studies), and the concrete ways that the urban environment is polluted are challenged, sustainability measures might turn out to be yet another ideal but unworkable policy. There is a real possibility that hitherto persistent and threatening contamination in cities can be reversed in the future. Yet, only if the contradiction between the continuous drive to amass economic wealth and the mechanisms that make environments unsustainable is confronted head-on, if lay and expert knowledge are connected, and if political mechanisms and actual perpetrators are singled out, may there be a chance of success.

Conclusion

This book addresses the questions of how and why the environmental balance in the large cities of highly economically developed countries has so frequently been disrupted. It draws attention to the persistent presence of air pollution over time, and challenges the economic trends that are implicated in its production. It incorporates relevant issues of knowledge formation and contests some of its limitations. This book studies the subject using specialized interdisciplinary views and explains it through a solid political-economy perspective.

Chapter 2 deals with the fragmentation of knowledge that has characterized the subject; Chapter 3 focuses on the political economy of the issue and develops the conceptual framework and an applicable model. Chapter 4 presents the actual making of a city into a wealthy world city and the concomitant rise of pollution. Chapters 5 to 7 give philosophical and political questions institutional, measurable and participatory empirical relevance. Chapter 5 looks into current monitored contamination and the incidence of ill health; Chapter 6 analyses ill health in relation to the health care system and to a number of socioeconomic indicators of poverty; Chapter 7 discusses environmental regulations and the political significance of spatial variation of ill health. In order to draw the social and natural dimensions closer together in explanatory terms, the book uses a combination of empirical, contextual and theoretical analyses. It applies them conceptually to explain the constancy of air pollution and the underestimation of its risk, and to assess in practice both the degree of environmental degradation and the extent of ill health in temporally and spatially specific contexts. Participation of a substantial population sample in comparative terms has facilitated the identification of the problem, and enabled better understanding of the conditions that make successful regulation of the environment a difficult, but possible, political enterprise.

The connectedness between social and biophysical dimensions, the difference that place makes in practice, the social commitments that dominate government regulations, the invisible but significant links between global and local economic growth – all are issues that are taken up at different stages in the book. Scientific knowledge that testifies to the extent, characteristics and mechanisms of air pollution and its effects on public health is used to tighten or promote a new, environmental legislation. Such is the case, for example, of the American and also the British Air Pollution Acts. These were sanctioned after recognition that the hideous episodes of urban air pollution in the 1940s and 1950s had caused a remarkable number of deaths. It should not be necessary, nor advisable, to re-experience extreme episodes similar to those of the 1950s in order to admit the unnatural presence of urban pollution, acknowledge its wide health risks, and promote more adequate controls.

Yet, society has forgotten the past and also chosen to ignore recent pollution episodes despite the fact that ample scientific research, regulatory controls, new emissions technology and improved management are in place to reduce air pollution. Our critical and also interdisciplinary framework forces an improved pattern of thought and it concludes on the politics of our historical amnesia. It assesses the institutions that represent inconsiderate economic growth and globalization and evaluates biophysical transformations with reference to particular geographical contexts. Written from a social sciences perspective, the book is firmly grounded on quantitative data, scientific knowledge, historical analysis, and, importantly, on a large and original database originated in reports by local residents. It also mentions views of public figures in the international city of Houston, the energy capital of the world.

2
The Knowledge on Urban Pollution and Health

Introduction

The absolute concentration of industrial and domestic pollutants has undeniably decreased since the 1950s, and its characteristics have noticeably changed in relation to the type of air contamination registered in the past. Yet, in the last decades, the insidious impact of lower levels of air pollution on health, and the damage that new pollutants and their combination may cause, have been the subject of increasing scientific recognition and have awoken public concern. Immediate and acute episodes of high concentrations of air pollution are not only a feature of the industrial past. It is only their degree of severity and the type of participating pollutants that make modern episodes of air pollution different from earlier ones. Despite cleaner air in many major cities, air pollution trends have thus, essentially, not been reversed.

The current state of the natural environment of many world cities seems to indicate that there have been deficiencies in the way that pollution in cities has been controlled and, moreover, in the ways that the environment has been understood. The sciences have competed with each other to become the explanatory tool for environmentally related problems. Whatever their shortcomings, it must, however, be stressed that both the social and natural sciences have contributed essential input for understanding changes in the chemical balance of the atmosphere, and relating human health disfunctions to this imbalance.

Yet, after the groundbreaking findings that identified the adverse health effects of lead on child health and intellectual function (for example, Needleman et al., 1979; Nyhan, 1985), and once atmospheric lead was dramatically reduced in most developed countries in the 1970s, the likely effect of other air-toxic materials was, notoriously, considered less problematic. However, there were still many airborne pollutants – equally or perhaps more dangerous than lead, and some have yet to be fully identified – which can put child health at serious risk. The severity of the problem has been raised in the literature and also in the media. Controversies have been raised in the epidemiology literature. The most common debate has been over the health effects of moderate to high levels of urban air pollution. Other problems have been raised over the health uncertainties of pollution levels below the official safety standards, the effects of long-term or chronic exposure to air pollution, and the insidious lasting effects of exposure (for example, Dockery et al., 1989; Keiding et al., 1995; Ostro et al., 1996; Villalbí et al., 1984). In the media, young children have been singled out because they are particularly exposed to pollution in the form of traffic emissions: 'Now we have pollution actually discharged straight into people's faces – especially small children sitting in push-chairs ' (Brace, 1994, p.4).

Older children may be equally exposed to car and industrial emissions when making their way to and from school. Moreover, children are likely to travel less than their parents, and therefore, child health can offer a more accurate study subject when measuring exposure to local air pollution.

Understanding the biophysical side of the relationship between urban air pollution and health is crucial and it requires the intervention of the natural disciplines, such as epidemiology, medicine, chemistry, toxicology and ecology (Bascom et al., 1996). Moreover, society has played a central role in this relationship in so far as it has enabled the conditions for pollution to proliferate. Buttel (1997) clarifies this point, arguing that environmental sociology is in some sense a materialist critique of mainstream sociology: 'Environmental sociology's agenda is, in part, to demonstrate that the biophysical environment matters in social life and that ostensibly social processes such as power relations and cultural systems have an underlying material basis or substratum' (p. 44).

But the approaches most often employed in the social perspective have focused on the sociology of health and the spatial perspectives of research. These have studied the inexorable changes caused to groups and individuals due both to exposure to pollutants and their to socioeconomic and demographic circumstances. Notwithstanding that the biophysical phenomenon of, for example, high concentrations of sulphur dioxide, of carbon monoxide and particles in the atmosphere might well have a strong political and economic origin, explanations of the phenomenon of rising pollution and related ill health have drawn principally on the natural sciences. As a consequence, research undertaken in the natural sciences has prevailed as the most common and authoritative source of knowledge to explain local as well as global biophysical transformations arising from toxic offensives launched on the environment.

Overall, social and biophysical disciplines have significantly contributed to our understanding of environmental degradation and its effects on human health, our evaluation of institutions for achieving clean air, and in suggesting ways to reduce pollution and ameliorate its consequences. The problem is that these scientific approaches have conceptualized the natural environment, human health and social questions as discrete entities. By doing so, they have inadvertently sidelined the importance of the political and economic structures in affecting those issues and have ignored any interactions between the natural mechanisms and the ways that society is organized. In the main, there has been an unfortunate fragmentation of knowledge, complemented both by an ahistorical perspective of biophysical transformations and by a limited economic and macro-political analysis. The result has been that crucial causal factors have been readily dropped from the analysis of health and pollutants. Decisions by governments on environmental policy, the degree of success of major economic processeses such as globalization, the strategies of international firms, and the character of public participation have all practically been overlooked. This book claims that such omission is due mainly to restrictive philosophical, political and methodological stances in those sciences dealing with the issue. Chapter 2 examines the main alternative explanations that have prevailed in the field and classifies them in three ways, namely, under the biophysical, the spatial and the sociology of health perspectives. It discusses their theoretical assumptions and methodological strands, and highlights those elements that are necessary for shaping a critical and comprehensive conceptual framework for assessing pollution and health in the context of globalization and world cities.

Alternative explanations and fragmentation of knowledge

There exists a wide range and massive quantity of literature on the topic of pollution and health. This literature has been classified in three different ways, each featuring different core scientific disciplines. Each group displays distinctive alternative approaches and specific types of information (see Figure 2.1). The first set of studies is based on the *biophysical approach* and refers to the literature in the natural sciences. This literature has aimed at producing very accurate descriptions of symptoms and the physiology of changes as expressed by the human body and in nature. Detailed explanations of disease draw particularly on epidemiology, toxicology and medicine; accounts of air pollution draw on environmental science. Information on health is collected to assess health risk, to measure individual levels of exposure in relation to concentrations of pollutants. The information gathered is then aggregated to allow the statistical analysis needed to reveal whether certain risk factors are associated with the prevalence of disease or mortality. The second group of studies has focused on the geographical distribution of disease and on the importance of place for providing clues to its distribution. This consists of the *spatial approach* and draws principally on the discipline of geography. Relevant mainstream debates in the

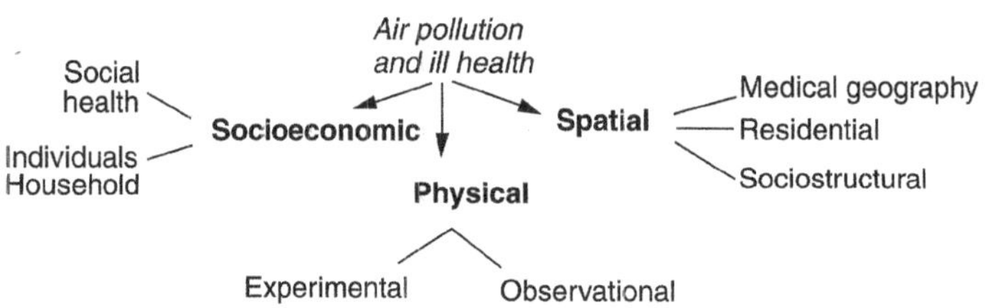

Figure 2.1 Analytical perspectives of air pollution and health

spatial approaches have been 'medical geography', 'residential location', and 'space matters'. The last perspective is represented by the *sociology of health approach*. Its input is important because it can fill out part of the incomplete picture depicted by the two previous, competent, although insufficient, approaches. The approaches comprising the sociological perspective consider the living conditions of individuals and groups. A most influential debate has been that on the relative effect of social class as defined by occupational status, and on the role of income. Another central theme focuses on the role of health care institutions and delivery of health services. A further debate from the sociological perspective has developed around the role of factors such as housing and family structure.

It should be stressed that in order to obtain a comprehensive understanding of the problem at hand, it is necessary also to assess the political and economic aspects of the problem, as well as to establish the interconnections between the different geographical local, regional and global levels. This analysis is undertaken in Chapter 3.

Crucial epidemiological evidence

Attention to problems caused by air pollution in cities was already paid in the nineteenth century. The remarkable incidence of death associated with air pollution represents – or at least, should represent – an eternal testimony of the extreme health damage that urban pollution can unleash. In our time, particularly during the last decades of the twentieth century, researchers in the natural sciences have produced consistent evidence firmly indicating that concentrations of air pollution can be harmful to human health. In the biophysical approach, a number of methodologies are used. For example, toxicology entails experimental research and intervention by the researcher such as examination for lung-tissue damage, animal exposure–response trials, and *in-vitro* studies. The other natural disciplines employ observational designs, and epidemiology is frequently complemented by varied demographic and socioeconomic information in some form of causal correlation. For example, the data usually include information on age, sex, weight, height, educational level, smoking history, occupational exposure and medical history, most of it requested as part of standard medical questionnaires. Epidemiologically standardized questionnaires on respiratory symptoms may follow one of the three basic types of observational design: cross-sectional, case-comparison and cohort-observational. Medical examination most often uses spirometric tests of pulmonary function. Environmental observational designs employ monitoring technology for pollutants, and atmospheric measurements of factors such as wind, humidity and temperature. They also use the satellite mapping technology on which national and regional control agencies widely rely for daily reporting on air conditions to the general public.

Cumulative evidence has emerged in particular from epidemiology, which has constructed a fundamental basis for establishing a causal association between air pollution and ill health. Due to space limitation and the need to review other bodies of literature, the epidemiological review is compactly presented, distinguishing three main lines of finding. One group has argued that air pollution may cause mortality, chronic and also temporary illness. Another strand indicates that children and infants are particularly vulnerable to pollutants, and, finally, there are the findings that show that levels of air pollution do not need to reach extreme levels in order to become harmful. A review is presented of the important findings and is followed by a discussion of the approach's limitations. Linking air pollution and mortality, a particularly useful study carried out in infants in the former Czechoslovakia between 1986 and 1988 observed substantial correlation between post-neonatal respiratory mortality and total suspended particle and sulphur dioxide levels (Bobak and Leon, 1992). This country had some of the highest levels of air pollution, producing the second highest annual emission of sulphur dioxide in Europe in 1987. In the Brazilian city of São Paulo which, unlike the rest of the country, is highly industrialized, respiratory disease is prevalent, and the most common cause of death for children under the age of four. In São Paulo, among people aged 60 and over, around 14 per cent of deaths are caused by respiratory problems, compared with the average 8.6 per cent for the rest of Brazil (Sunnucks and Osorio, 1992).

However, morbidity, rather than mortality, seems likely to be a more sensitive measure in studies of relatively long-term and high, but not extreme, levels of air pollution (Bach, 1972; Collins et al., 1971; Davies, 1994). Rates of respiratory symptoms were found to be especially high among children in cities with high particulate pollution (Dockery et al., 1989; Forsberg et al., 1997; Saraclar et al., 1998). Further, a 20 per cent higher risk of children developing respiratory problems in Mexico City was registered when they were exposed to ozone pollution peak, > 0.13 ppm, for two consecutive days in 1988 (Anderson et al., 1997; Romieu et al., 1993). Highly industrialized Mexico City has frequently been identified as one of the most polluted areas in the world, with almost half of all babies registering dangerous levels of lead in the blood (Elsom, 1996; Schteingart, 1989). Mexico City is encircled by mountains and subjected to prolonged periods of light winds and strong temperature inversions which trap the pollutants over the city and then cause illness. It was found that chronic cough, bronchitis and chest illness in both children and adults (controlling for household variables such as parental smoking, type of cooking stove, and history of respiratory illness) were found to be positively associated with all measures of urban particulate pollution (Schwartz et al., 1991). Similarly, a study in Israel similarly showed how greater prevalence of most respiratory symptoms was found among children who lived in more polluted than in less polluted areas, having controlled for household condition (although these were also important) (Goren et al., 1990).

Kinney et al. (1996) has provided strong evidence that children attending six summer camps in the US were exposed to concentrations of ozone and definitely experienced decreases in lung function of the kind demonstrated in laboratory studies. Some controversial conclusions regarding exposure to ozone have been broached. Steadman et al. (1997) estimate that the additional hospital admissions in the UK for respiratory disease attributable to ozone levels above the threshold of the recommended maximum concentration of 50 ppb were very small during the summers of 1993 to 1995. On the other hand, Spix (1997) contends that it is likely that short-term effects such as hospital admission are only part of the profile of the effects of ozone. Individuals and populations seem to be able to adapt to higher levels of ozone in terms of short-term effects on lung-function paramaters. However, animal experiments have shown that long-term exposure causes lasting damage to the lung tissue. Despite the controversy, there has been growing proof that high levels of ozone exacerbate pre-existing respiratory disease and cause increases in emergency attendance, hospital admissions and mortality (Ostro et al., 1996). Substantial studies had observed that the effects of acute exposure to PM-10 cause a decline in lung function. These are often accompanied by symptoms such as chest pain, coughing, nausea and pulmonary congestion, a decline which may persist for weeks after a single pollution episode (Dockery et al., 1989; Lewis et al., 1998; Lueunberg et al., 2000; Ransom and Pope III, 1992; Schwartz et al., 1993). Also emergency room visits (Castellsague et al., 1995) and asthma symptoms have been usual (Ostro et al., 1995; Peters et al., 1996). Analysis of daily urgent hospital admission for respiratory illness in Montreal, Canada also found an association with PM-10 (Delfino et al., 1994). High levels of school absenteeism in Utah Valley, US were associated with exposure to high concentrations of PM-10 emitted by an integrated steel mill (Ransom and Pope III, 1992; see also Pope III et al., 1991). In Barcelona, Spain, hospital admissions increased when air pollution was very high, and 48 per cent of all patients admitted were suffering from asthma or bronchospasm emergencies (Villalbí et al., 1984). In the Chilean capital, Santiago, there is evidence to suggest that one out of three children suffers from bronchitis, and visits to doctors for the treatment of respiratory problems are higher than the annual average worldwide (Sunnucks and Osorio, 1992).

A large number of studies suggest that even concentrations of pollutants lower than those given as guidelines in many countries may increase the incidence of respiratory illness. A large-scale study carried out in the 1970s significantly revealed a close correlation between low levels of air pollution and various respiratory indices in children from eight different European countries: Czechoslovakia, Denmark, Greece, the Netherlands, Poland, Romania, Spain and Yugoslavia (WHO, 1979). Another useful three-year study of asthma and pollution in Helsinki has shown that, among children, exposure to atmospheric ozone and nitric acid was significantly correlated with admissions to hospital even though the levels of these pollutants were fairly low (Pönkä, 1991). Changes in the biochemistry of the lungs have been documented at ozone levels well below international safety limits. Read and Read (1991) point out that long-term exposure to ozone for 6–7 hours at relatively low concentrations, that is 0.08 ppm, has been found significantly to reduce lung function in normal, healthy people during periods of moderate exercise. Similar investigations were carried out in Italy (Forastieri et al., 1992), Spain (Mallol and Nogues, 1991), Switzerland (Rutishauser et al., 1990), Mexico (Romieu et al., 1993), France (Quénel and Médina, 1993; Boussin et al., 1989), and Finland (Jaakkola et al., 1990).

These show that the frequency of respiratory symptoms in children is correlated with levels of contamination which usually do not exceed any established 'safety limits'. Kühni and Sennhauser (1995) in Switzerland, and Keiding et al. (1995) in Denmark carried out independent studies on school children. Yet the two groups concluded that even moderate to low ambient nitrogen dioxide and nitric oxide, as indicators of traffic-related pollution, appeared to be related to an increased prevalence of chronic respiratory symptoms such as wheeze at night and restrictions in daily activities.

Inadequacies in the biophysical approach

The problem that arises here is that the natural sciences are inadequate on their own to understand air pollution and ill health in contemporary society, and unless the limitations of the biophysical perspective are highlighted, the benefits to be gained from this vigorous scientific research may well be lost. It is not our purpose to criticize the necessary information produced by the discipline. What is challenged is the application of positivist methods to explain and produce a more accurate understanding of the problem of persistent contamination and ill health in cities. It requires an interdisciplinary and political-economy analysis of this complex environmental and medical question and some competent works have rightly stressed these serious limitations. According to Lewontin and Levins (1997), human beings clearly have certain biological properties of anatomy and physiology that both constrain and enable them; no historical contingency or change in consciousness can remove those factors. But at the same time, the central nervous system of human beings combined with their organs of speech and manipulative hands, lead to the formation of social structures that produce historical forms and transformations of those needs. Without siding with the scientific pessimism of Ulrich Beck (see for example, Beck, 1997), some useful insights can be extracted from his writings. He stresses that there exists a danger that an environmental discussion conducted exclusively in chemical, biological and technological terms will inadvertently include human beings in the picture only as organic material. This attitude runs the risk of atrophying the subject into a discussion of nature without people, without asking about matters of social and cultural significance (Beck, 1993). Rather than operating on a purely biological way, a social system fundamentally operates in relation to environmental risks. Beck (1997) argues that 'Hazards are created by industries, externalized by business, individualized by the legal system and trivialized by politics' (p. 26).

A considerable difficulty for our analysis is that the natural sciences reduce the focus to objects and events, trying to explain properties of the complex whole – which is what we want to understand – simply in terms of the units of which it is composed. This positivist epistemology accounts for observed phenomena by the development of laws and generalizations based on empirical regularities. A fundamental requirement that substantiates the biophysical explanations is that observations and interventions are supposed to be independent of one another, as if they existed, in reality, in isolation. The story of the drunk who tried to discover the causes of his drunkenness by using such methods (Sayer, 1992) helps to clarify this separation and also to highlight the reductionism embedded in it. On Monday, he had whiskey and soda; on Tuesday, gin and soda; on Wednesday, vodka and soda; and on other nights, when he stayed sober, he drank nothing. By looking for the common factor in the drinking pattern for the nights when he got drunk, he decided the soda water was responsible. Now, the drunk might possibly have chosen alcohol as the common factor and hence as the cause. However, using our methodology of critical realism, as specified in Chapter 1, what gives such an inference credibility is not merely that alcohol was a common factor but the pre-existing knowledge that it has a mechanism capable of inducing drunkenness. The implications for right deductions provided by this seemingly trivial example are certainly far-fetched even though for many applications in social science the explanatory situation is more complex. There is not one but several equivalents of the soda water and it is much more difficult to separate soda water and alcohol. While looking for differences between situations seems sensible, such an approach is inconclusive in causal terms because there can be regularity without causality. For example, to focus exclusively on the behaviour of the drunk runs the risk of hiding important preconditions for drunkenness, which rest, instead, with the collective trends of a society at a particular time in history.

Now, extending the reductionist linear model of correlation and association of similarity but including some socioeconomic indicators such as social class and educational levels, or adding geographical differences, as many epidemiological studies have done, does not constitute a sufficient break with basic deficiencies of traditional scientific methods; for, while the biomedical model has been extended backwards, it has not been extended far enough to include societal, structural variables (Jones and Moon, 1987). For example, children living in deprivation are vulnerable to the effects of environmental lead on two counts. First, these children run the same risk as other urban children when exposed to the properties of atmospheric lead that cause neurological and behavioural damage (Agency for Toxic Substances and Disease Registry, 1988; see Baghurst et al., 1992; Centers for Disease Control, 1988; Needleman et al., 1979; US EPA, 1986). Second, because of their poverty, children in low-income households are at higher risk of experiencing the consequences of lead-contamination, due to factors such as precarious health care, residence near either lead-contaminated areas or emitting abandoned sites, or major roads, and inadequate diet (for example, Mahaffey et al., 1982).[1]

Of additional importance is the fact that the biophysical, as well as the sociological, approach described above perceives the phenomenon of environmental ill health through an *ahistorical* lens. That is, a more realistic and historical perception of hazardous concentrations of lead would incorporate the fact that lead is found in the environment primarily because it is used ubiquitously in car batteries, paint and fuel. Therefore, lead represents an essential component for industry and energy in modern society. Decision-making on the use of this metal has been driven not only by technical factors, but also by the political priorities of leaders to encourage, reduce, or ban it. In summary, all human organisms vary, both in response to the intricate patterns of environment and also because of their own internal dynamics. For most medical and epidemiological research, external variation is a nuisance, and much ingenuity goes into removing the variation experimentally or statistically in order to detect average or main effects (Benton, 1994). However, variation between organisms within a species is the necessary ingredient for understanding the processes of evolution, by natural selection and as topic of interest in its own right (Lewontin and Levins, 1996). The example of the drunk illustrates that neither common nor distinguishing properties need be causally relevant. By usually ignoring qualitative and also changing political and economic priorities over time, and by abstracting from variation in their contexts, even when these are often linked and internally related, large parts of the sciences have overlooked precisely these very things that we are interested in from an explanatory point of view – interdependence, connection and historical emergence.

The geograpical perspectives

Our spatial approach draws on geographic writings which address the issue of ill health, the environment and social structures. Three main perspectives are most relevant here: medical geography; residential location; and the social construction of space.

In the first approach, *medical geography* is the discipline that has typically established the connection between society and disease distribution following spatial patterns. 'Disease ecology' is the main tradition in medical geography that attempts to elucidate the social and environmental causes of ill health following patterns of spatial distribution of disease (Jones and Moon, 1987). Such work is closely allied to epidemiology and may compare the health and disease of groups defined by household composition, inheritance, individual behaviour and environment (Morris, 1975). Geographical and correlation health studies have employed large-scale techniques over extended territory (for example, McGlashan, 1972; Pyle, 1979; WHO, 1979; World Resources Institute, 1988). Early surveys such as those conducted by Howe (1975) and Howe (1963) concluded that the most striking feature of the regional distribution of chronic bronchitis is that consequent mortality in the UK was about thirty times greater than in the US and five or six times greater than in most of Western Europe. While epidemiology focuses on individuals' risks in relation to a causal factor, medical geography draws on wide census information to trace the distribution of ill health. However, the very scale on which these geographical surveys take place limits their value in the study of risk factors that are widely distributed. Therefore, epidemiologists have emphasized that the implication is that they are unable to detect risks associated with environmental factors that are spatially localized (Cuzick and Elliot, 1992). One of the earliest such medical geographic studies, and still the most famous example of associative studies on the small scale, is that of John Snow in the 1850s. Snow, who was a physician, plotted the distribution of cholera deaths in London and discovered that the vast majority had lived in one area and drunk water from one specific pump, under the supervision of the Southwark and Vauxhall Company's water supply (Blaxter, 1975; Eyles and Woods 1983). John Snow's single-dot distribution map of cholera deaths in London and Howe's (1963) early maps illustrating the distribution of a number of diseases in the UK are often cited as models which proved the method's usefulness (McGlashan, 1972). From his maps of bronchitis mortality, Howe deduced that regions of high mortality in the UK corresponded to areas of dense industrial population. Lunn et al. (1970) and Girt (1972) pointed out that air pollution, which was seen as a characteristic of UK industrial cities, was clearly implicated as an important factor in childhood respiratory disease and in chronic bronchitis in general.

Importantly, later studies of cancer mortality suggest, first, that the use of large areas may mask the presence of aetiologic factors in the local environment, as affirmed by the finding of increased incidence of childhood cancer. This discovery was possible because of small-scale epidemiological studies in the vicinity of nuclear installations (see Gardner, 1989, 1991; Gardner et al., 1990; Shleien et al., 1991), and because the arbitrary selection of 'boundaries' to categorize proximity – where the non-exposed are misclassified as exposed – can be highly influential on the results obtained (Hatch, 1992). Small-scale geographical associations were also found between mortality from cancer of the respiratory tract and heart disease, and regional air pollution samples in Houston, US (MacDonald, 1976). Caprio et al. (1975) showed that rates of excessive lead absorption in children are related to proximity to urban roadways and traffic volumes. This type of study has obvious importance and provides clues for further investigations because location is contingent on other factors.

The next spatial perspective that addresses environmental degradation and ill health is *residential location* and focuses on the characteristics of living areas which might induce illness. Much research has been carried out on the association between features of local area of residence and health damage (see, for example, Blaxter, 1990; Britton et al., 1990; Burr et al., 1997; Sobral, 1989). Although ostensibly about geographical variations in health, many of these studies are not, *per se* about the role of areas in influencing health. Rather they use areas as vehicles for exploring hypotheses about the role of biophysical exposure or material deprivation in the aetiology of ill health (Macintyre et al., 1993).

A main premise in this literature is that over and above individual-level attributes of deprivation, people of low socioeconomic status may have poorer health because they tend to live in areas which in some ways are health damaging (Blaxter, 1983; Haan et al., 1987). Colley and Reid (1970) suggested a consistent class gradient of frequency of chest conditions with air pollution in urban and rural environments in England and Wales for the children of social class IV and V only (see below). Girt (1972) claimed that people with low-income and low occupational status tend to live in the most polluted areas of town. The reason was not only because of the proximity of the factory and poor-quality cheaper housing, but also because of the spatial concentration of small working-class houses – featuring poor construction, damp and overcrowding. Sobral (1989) argues that in São Paulo, the pre-eminent Brazilian industrial city, the prevalence rates of respiratory diseases were higher in the areas with higher pollution levels, particularly in the slums.

However, the argument that poverty and illness tend to cluster and to reinforce each other confuses causality for two reasons. First, air pollution contaminates not only low-income areas. This is due to a number of factors: to the power of pollutants to affect areas far from sources of emission because they travel easily; to mix and create new pollutants with new properties; and to the widespread character of polluting activity, such as motorization. Access to better living conditions for low-income populations, including adequate diet, medicine and housing, must be a social priority. However, only improving the living conditions of individuals in response to the health problems posed by air contamination in local areas is a very partial 'solution'. The contribution of these factors to immediate better quality of life is important. However, solutions of this type alone may only mitigate, rather than eradicate, air pollution and localized health problems, overlooking the structural origin of the limited access to health care, the pervasive character of pollutants, and the increasing amount and wide distribution of emissions in large cities. Second, a debate founded only on ecological and geographical features of residential area ignores the fact that social relations of production create and foster such geographical variations, some for their own ends. While each geographical area, such as those contained in global cities, is unique, it also fits a pattern of visible and less visible urban environmental degradation worldwide.

While not denying that environmental hazards can behave democratically – any population might fall prey to environmental risks – as this research shows, a more 'just' distribution of pollutants will not make contamination any more bearable or acceptable to the public. The addition of the political and economic analysis, and the discussion of past events as presented in the next chapter, will enable those geographic patterns that are indicative of a contradiction between growth and the environment as seen to be highlighted.

Our last spatial perspective is the *social construction of space*. Contrary to the previous configurations, this geographical literature does not focus on the relation between environment and health but develops instead a conceptualization of the relations that are found over space. The approach is useful for explaining the connection between local industrial pollution and residents' illness, connecting between the macro-structures of society at large and the characteristics of local places. Whilst geographical views of the 1960s claimed a purely spatial or geographical world, devoid of substance or content, the 1970s saw the underestimation of geography as distance and in terms of local variation and uniqueness (Duncan, 1989; Massey, 1996, 1987; Savage and Duncan, 1990). However, there are no such things as purely spatial processes; there are only particular social processes operating over space. Neither is the spatial a pure social construct that can deprive geography of its spatial role. Space should not be viewed as an absolute entity somehow separated from the material objects located 'within' it. But it is also the case that space cannot be merely reduced to such objects (Urry, 1987). Space is a social construct but social relations are also constructed over space, and that makes a difference. For the theorists indebted to this approach, the economic reorganization of society provides an important reference for the construction of contemporary space. According to Massey (1996):

> It is necessary to move beyond the characterization of globalization in terms of speed-up, instantaneous communication and constant global flows to imagine the process in terms of the spatial reorganization or social relations, where those social relations are full of power and meaning, and where social groups are differentially placed in relation to this reorganization.
>
> (p. 121)

An assessment of geographical patterns of ill health is crucially necessary in order to reveal the prevalence of disease and possible environmental associations. However, structures and mechanisms underlying these particular geographic variations are more important from a causal point of view. The fact that social processes take place over space, the facts of distance, or closeness, of the individual atmospheric character of specific places – all these components are essential to the operation of ecological reactions and health changes. 'Just as there are no purely spatial processes, neither are there any non-spatial social processes' (Massey, 1987, p. 52). Duncan (1989) points out that, clearly, it is not spatial location *per se* that accounts for variations, for spatial relations are still secondary and contingent, even if primary, generative causal mechanisms are spatially bounded.

This approach highlights the fact that 'space' is not a passive surface on to which the relations of production are mapped, nor yet simply a negative constraint (in the sense, for instance, of distance to be crossed). Spatiality is an integral and active condition. This approach attempts to clarify the way in which spatial inequality is both produced and used by firms in their search for conditions favourable for continued capital accumulation (Massey and Meegan, 1985). It draws on the premise that production is distributed and organized systematically over space and that, in a fundamentally capitalist society, the system's rationale is the pursuit of profitable growth. Production, spatial form and spatial strategy can be active elements of accumulation. Massey (1987) saw the challenge in the construction of an approach which is neither empirical detailed description, nor a mechanistic Marxist insensitivity, and in the recognition of any specificity within the grander historical movements of capitalist societies.

This is similar to our story of the drunk who 'by looking for the common factor in the drinking pattern when he got drunk, decides the soda water was the cause'. Spatial location in itself does not explain the incidence of ill health because, as mentioned previously, place *per se* cannot make the events happen. Even though concrete studies may not be interested in spatial form *per se*, as Sayer (1992) argues, it must be taken into account if the contingencies of the concrete and the differences they make to outcomes are to be understood. The central role of spatial relations can be highlighted by paraphrasing Pratt: 'space is no longer just "context" or "contingency"; it is – in its re-conceptualized social-spatial mode – something that is constructed by and, in turn, constructs, social conditions' (Pratt, 1994, p. 205).

In empirical research on concrete objects and processes, analysis of the situation regarding space involves investigating the actual workings and effects of mechanisms in contingent circumstances. Sayer (1992) argues that a considerable amount of social research is weakened by a largely unnoticed scrambling of causal forms. At worst, the degree of abstraction from the actual forms in which objects relate is such that the processes by which mechanisms produce their effects are simply obscured and become lost in an aggregate, 'de-spatialized', statistical soup. He argues that 'The less explanations of actual events take account of the contingencies of spatial form, the less concrete they can claim to be' (p. 151).

Not surprisingly, regularities are at best transient and spatially limited. Depending on the nature of the constituents, their spatial relations may make a crucial difference. Spatial contingency will be influential on how social and biophysical processes work and what forms result. For example, if manufacturing oil facilities are placed near residential areas, sulphuric emissions will most probably adversely affect children living nearby whereas children living at a distance will not be affected in the same way.

In summary, it is invalid to ignore the fact that biophysical mechanisms, policy implementations, ecological transformations, and social relations take place over a geographically differentiated world. None the less, to say that geography matters is not to say that space alters the processes themselves, although spatial relations crucially contribute to empirical variations. Spatial patterns are not independent of social and biological processes, nor does space determine behaviour. Social and spatial changes are connected to each other. Therefore, the space as socially constructed, rather than only as the place for geographical distribution, is an essential dimension of the study of ill health and air pollution because the two also connect to economic and policy trends.

The sociology of health deprivation and its indicators

The last approach to environmental degradation and ill health uses the parameters of the sociology of health. One notable feature is that the distribution of disease in society is not conceived as something which flows only from the innate properties of individuals and involuntary exposure to toxins, but rather as emanating from the institutional organization of society and individuals' socioeconomic conditions. A remarkably good literature on the state of health in relation to social inequality and exclusion was produced during the 1970s and 1980s. The US National Center for Health Statistics estimated that in 1980, more than 29 million Americans (14.9 per cent of the population) were in a low per capita household-income category. In this group, there were twice as many people limited in their activities because of chronic health problems (29.3 per cent) as was the average for members of households of all incomes (14.5 per cent), and over three times (8.7 per cent) as many as the percentage in households with annual incomes above the federal poverty level (Dougherty, 1988). It is striking that similar inequalities to those pointed at in the above earlier literature continue to prevail in the developed countries. Deprivation and ill health were still found to be associated in the UK in the 1990s (Wihill, 1994).

Two main foci can be distinguished in this literature, often interconnected. One group refers to the role of social class in determining ill health and health care provision, and the importance of other parameters of social inequality such as household income, housing, family structure, and also individual habits and lifestyle, for ill health.

The social class view has emphasized that there is no longer much argument that, in general, those groups of low socioeconomic status, poor urban environment, unskilled occupation and under-resources are likely to suffer earlier from greater morbidity. This relationship has been documented for the UK, where there is a long tradition of classifying the population according to occupational positions (Blaxter, 1981; Fox et al., 1985). Social class is a summary indicator of social inequality. It has its origin in social institutions which are outside the individual's control. The basis for ascribing social class has been the level of occupational skill: I – Professional; II – Intermediate; IIIN – Skilled non-manual; IIIM – Skilled manual; IV – Partly skilled; and V – Unskilled (Office of Population Censuses and Surveys, 1973). Earlier large-scale studies by Spence et al. (1954) and Dawson et al. (1969) of medical consultation figures assembled by occupational class and by health condition revealed high levels of illness, incidence of severe respiratory disease, total days of illness, and medical consultation for serious respiratory disease that were at conspicuously higher rates for the lower occupational classes. High proportions of severely asthmatic children were found in semi- and unskilled manual households. The use of better indices of social deprivation, including geographical measurements such as mechanisms for studying areas by postcode or municipal wards, has obviously enhanced the information, but without altering the overall result.

This approach has found that mortality tends to rise inversely with falling occupational rank or status for both sexes and at all ages. Availability of medical care in poor industrial areas in the UK decreases proportionately to need (Black et al., 1982). From this perspective health inequality is a measure of the social environment and its capacity to generate inequalities of welfare and survival (Hart, N. 1986). Hart, J.T. (1975) has summed up these trends as the inverse care law: that the availability of good medical care tends to vary inversely with the need of the population served. A main argument is that social class differences in morbidity and mortality are very pronounced, with a class 'gradient' and poverty being historically associated with respiratory ailments and vice versa.[2]

Health inequalities in England specifically among children emerge as particularly marked among different social groups (see, for example, Blaxter, 1975 and Hart, 1986). At birth and during the first month of life the risk of death in households of unskilled workers is double that in professional households. For the next eleven months of a child's life this ratio widens still further: for the death of every one male infant of professional parents, almost two deaths among children of skilled manual workers and three among children of unskilled manual workers are to be expected. Among females the ratios are even greater. Socioeconomic inequalities encountered during childhood persist so that ill health is worse, and death rates higher at every stage of life the poorer the person is. There is evidence that social class trends in the experience of chronic illness, or in the proportion of people who assess their own health as 'poor', are steeper than class differences in mortality (Blaxter, 1990). This is the case even when mortality rather than morbidity rates have been considered as the best available indicators of the health of different social or, more strictly, occupational groups (Black et al., 1982).

Accounting for occupational social class has uncovered substantial differences in health. There are, though, a number of difficulties because social class in the strict sense of occupational status itself cannot cause disease, but acts as a marker for differences between groups of people, including housing, income and education, use of health services, diet and incidence of stressful life events (Golding, 1986): when evaluating the effects of air pollution on health, it is imperative to consider a full range of factors. Indeed, as Black et al. (1982) have emphasized, significant as social class is, it is not a sufficient explanation. Social class differences do not explain anything. The reification of the analytical construct of class into something substantive in itself diverts attention from the ways in which the social and material circumstances of individuals provide different constellations of risk (Oakley, 1992). This limitation can be partly overcome by examining the specific conditions in the household and their relationship to child health.

Also, historically, occupational status has been defined in a gender-biased manner. Most precisely, the male head of the household has been selected as the principal indicator of social class on the basis of his occupation, but women and children are only assigned occupational-based class as a result of their membership of a household headed by a man. Therefore, while social class might be predictive of life chances, it refers to the lives of men and not necessarily to all members of the household (Oakley, 1992). This has obvious ramifications when attempting to examine the health of children in, for example, single-parent households as well as in high-status social-class homes located in the surroundings of industrial development. In the context of the current book, which focuses on the health of children in a variety of residential locations, there are therefore considerable limitations to attributing causality only in relation to social class.

The second outlook focuses on the dominant influence that distinctive household socioeconomic factors exert on health status. Townsend et al. (1988) have emphasized, however, that socioeconomic variables are difficult to measure and trends in inequalities in the distribution of income and wealth, for example, cannot yet be related to indicators of health, except indirectly. For example, in all classes, owner-occupiers have lower mortality than those paying rent. Family structure, although only rarely included in present studies, has attracted the attention of some researchers. For example, Blackman et al. (1989) explored the health of the children of single-parent households compared with two-parent households, but their findings claimed no significant difference. In the UK the growth of absolute levels of resources, the spread of employer welfare benefits and of social service benefits, as well as the increase of owner-occupation among the working class makes a measure of resources all the more important. Income is a nearer measure. Household income may allow, or disallow, access to medical care and to other basic components of good health. Though household income is not available in clinical data, there is strong evidence that people on low incomes suffer more ill health. In the US, the health status of the poor is far below that of other income groups (see, for example, Hicks et al., 1989; Miller et al., 1985).[3] Turning to Canada, Dougherty, G.E. (1986) calculated that the overall aetiologic fraction of poverty, or attributable risk percentage of the mortality rate, was 30 per cent. That is, 30 per cent of the child mortality in low-income groups is attributable to their poverty. Jolly (1990) ascertained a significant increased mortality rate for most age groups, and for the younger age in particular, for the lowest income group. The Australian case corroborates the above findings where, moreover, the rate of prenatal mortality has historically been very low, among the lowest in the world. It remained the case in the 1980s that perinatal mortality was worse for single mothers (who are generally poorer) and for households at the lower end of the socioeconomic status scale (Hicks et al., 1989).

Further, Harlap et al. (1973) and Leeder et al. (1976) claimed that the risk for infants and children of becoming ill increases as the number of children in the household rises because infections are more likely to be spread within larger households. Another explanation offered for increased levels of ill health is that mothers of large households are less likely to have their children immunized against measles or pertussis (Butler and Golding, 1986). However, large sibship seems to have positive effects in the occurrence of allergic diseases of younger children. Housing has been an important focus in the sociological literature. For example, in the UK, inequalities of housing and health are still linked in spite of the dramatic expansion of state housing after the Second World War (Black et al., 1982). Council tenants were found to be 'less healthy' than owner-occupiers, and their unfavourable health records were primarily connected with material deprivation (Townsend, 1979). In West

Belfast, marked differences were found between the self-reported health of tenants living in 'good' as opposed to 'bad' council housing areas, allowing for social class, age, smoking and drinking habits (Blackman et al., 1989). However, Golding (1986) argues that living in new, centrally heated houses is not always healthier than living in older, un-modernized dwellings. Indeed, two earlier studies, one in America (Spivey and Radford, 1979) and the other in South Wales (Yarnell and Leger, 1977) have shown a higher rate of illness among children residing in newer public council housing compared with those in older private housing. Not surprisingly, the condition of dampness in the house was found to influence the level of respiratory/bronchial symptoms, and greater incidence of headaches, diarrhoea and aches and pains have been reported among children in damp dwellings than in dry ones (Forsberg et al., 1997; Hart, 1986).[4]

Occupational class and the other socioeconomic indicators previously mentioned definitively affect residents' health status. However, such determinants leave out other essential contributors to ill health. For example, the chemical condition of the outdoor air due to the location of industrial plants may not be directly related to the social class of the dwellers of nearby households, as this book shows. Exclusion of environmental and biophysical issues from investigation poses the typical problem of social determinism and socioeconomic reductionism, which under the guise of a methodological division of scientific labour separates the concerns of social and of natural scientists (Benton, 1994). Thus, while the analysis of people's occupational social class and of other factors such as housing and income may satisfactorily explain social inequality and health at the micro-structural level of sociological enquiry, those factors alone are insufficient to account for other commonly found situations. Similarly, Dickens (1992) suggested that 'Sociology constructs itself "as a watertight discipline largely by creating an impermeable division between itself and the natural sciences"' (p. 19).

Conclusion

This overview of the biophysical approach has illustrated that the human body and the natural environment possess their own peculiar physiology, ecological mechanisms and chemical characteristics. Even though society has frequently interfered with, and attempted to change, the course of nature, these self-regulating features are entirely independent of society and remain so. Therefore, ill health that relates to air pollution must be understood at its face value. The information that the natural sciences provides is necessary, for it must lead to further exploration of the type that will ensure political application. It is apparent that a combination of phenomena is taking place. The literature has indicated that disruption of the environmental balance in cities may cause symptoms of ill health when inhabitants are exposed to air contamination. It also tells us that indicators of social class, deficient access to health care and poor living conditions are all factors that trigger ill health. The characteristics of local place, and the social relations that have contributed to define what occurs in those places, play important roles in the status of residents' health. Yet, while this research is firmly rooted in the social sciences, it nevertheless depends heavily on the natural sciences; none the less, it does not rely on their explanatory power. This chapter has acknowledged the limitations of the methodology and analysis of available scientific studies, to which the present argument is nevertheless heavily indebted.

A number of problems have emerged from the review of current scientific approaches of the relationship between air pollution and child health because every line of explanation was incomplete in itself. The preoccupations of the sciences have been reductionist because their main interests have been either the biophysical, the social, the spatial, the local, or just the global. Each of these aspects has played a more prominent role than theoretical abstraction of the hidden processes that promote the relationship among those various dimensions. Alone, neither the analyses of the biophysical sciences nor those of the social sciences are sufficient to explain the relationship between air pollution and ill health. While the biophysical approach has not given enough attention to social, economic and political dimensions, the sociology of health tends to go to the opposite extreme, barely acknowledging the independent reality of nature – and that of the environment and the human organism in particular. They each explain the properties of complex wholes in terms of the parts or units of which they are composed and overlook political processes that affect the context of transformations of nature. Unfortunately, researchers from different scientific fields have mostly failed to enquire into essential matters that connect the aspects of the problem.

The sociological approach certainly recognizes the effect of social institutions such as social class and occupation, but tends to ignore the overall effect of the process of neo-liberal economic growth. Lewontin and Levins (1997) help to highlight this problem. They explain that, at best, liberal thought attempts to combine the biological and the social in a statistical model that assigns relative weights to the two, allowing for some component of interaction between them. But the division of causality between the distinct biological and social causes, which may then interact, misses the real nature of their codetermination. In this sense, emphasize Jones and Moon (1987), we need the macro-gaze of social structure and not just the micro-gaze of biological individualism to provide an adequate explanation of ill health as a social phenomenon. It is necessary not to reduce the social to the individual or to explain disease by only biological variations, but to do the reverse, that is to place and relate the individual and their biology in social and political context.

The chapter has shown that much of the work in the field of environmental degradation and health has been concerned with establishing universal, regular empirical connections. The interrelationship between biophysical and social events has been dimmed. Causal links between economic changes that take place over time and the spatial and social variations of human ill health and the extent of environmental pollution found at a particular place and time have remained not only unresearched but, more often, unacknowledged. While description and quantification are necessary, these cannot replace causal explanations. Empiricist methods as tools to approach ill health associated with air pollution by measuring the extent of the events and their associations, while helpful, are not always sufficient. Positivism should not be equated with quantification, which can be applied in any approach (Johnston, 1994).

Essential as natural science investigations and identification of geographical patterns might be, actual connections and interactions between objects have often been recorded in aggregates using their positivist methods alone, without any reminder of their links to some basic phenomena of current society such as economic growth. While explanation in positivism is often considered to be synonymous with establishing the causes of events, there remains fierce controversy over whether or not the real world actually behaves in such a manner (Jones and Moon, 1987). Positivism, however, should not be equated with quantification which can be applied in any approach (Johnston, 1994). The weak concern with the conceptualization of the linkages between social and biophysical realities emerges in large measure as a consequence of the observation that either air pollution or ill health is assumed to be unproblematic, their description a minor preliminary to the business of science. In this way, the explanation of pollution in highly economically developed cities remains incomplete and fetishistic.

The social mechanisms that make air pollution in cities something endured as an accepted risk factor, particularly after the literature has widely acknowledged its deleterious health effects, have been less addressed and researched than other aspects of the association. Whereas conventional research and institutional views of the problem typically separate the various components of the relationship, the challenge lies with addressing it as a multi-layered configuration and with opting for a critical and political explanation of the problem. The analytical perspectives and the knowledge produced in the three theoretical groups have furnished our understanding with highly informative data, methodological tools and some material for connecting the prevailing fragmented disciplinary conceptions. However, an analysis informed by these elements alone remains unconnected overall because it still lacks a wider framework of the political and economic relations that, at the local, regional and global levels, urban contamination make possible.

It is argued that this deficiency of employing any single disciplinary analysis in isolation shows a conceptual discontinuity among the biophysical and social sides of health, air pollution and the dominant social structures in major cities. This obstacle has contributed to emphasizing an ahistorical interpretation of the relationship. In fact, the commonly ascribed nature/society dichotomy muddies our analysis because our modern understanding of nature and development is embedded in the division between natural and unnatural, rather than in reality (Norgaard, 1994). Typically, natural and social scientists alike do not regard the presence of macro-political and economic processes which are historically determined as in any way problematic. The next chapter addresses precisely the role of economic growth and its impact on the environment.

Notes

1. For further information on the effects of lead on child health see Brown et al. (1990); Caprio et al. (1975); Needleman et al. (1979); Schneider and Lavenhar (1986); Wilson (1983).
2. In UK areas with most sickness and death, general practitioners have more work, larger lists, less hospital support and inherit more clinically ineffective traditions of consultation than in the healthiest areas; and hospital doctors shoulder heavier case-loads with less staff and equipment, more obsolete buildings and suffer recurrent crises in the availability of beds and replacement staff (Hart, J.T., 1975).
3. In the US, mortality rates are calculated in combination for racial, sex and age groups. It is therefore impossible to derive the actual incidence of death for the lower-income groups alone. None the less, the overall mortality rate is higher for Blacks and other minority groups, most of whom are concentrated in the poorer stratum, than for Whites (Dougherty, C.J., 1988).
4. Mould growth could be responsible for the significantly worse health of children in damp houses. Spores germinating under moist conditions may enter the respiratory tract, causing bronchial and asthmatic symptoms including fever, tiredness and lethargy. Children then have less resistance to allergens and are more vulnerable. In addition, allergic reactions may occur to the house dust mites and storage mites that multiply in damp conditions. The mycotoxins, or mould given off by fungi, may get into the mouth or nose and be swallowed, causing stomach upsets as well (Hart, 1986).

3

Theory and Practice of Growth and Degradation of Nature

Introduction

Chapter 2 identified the main theoretical paths that define, measure and analyse those factors that are constituents of the relation of air pollution and ill health. Social theory and the natural sciences can certainly continue to make major contributions to the understanding of environmental problems, but the danger is that their comparatively distinct disciplinary compartments will remain unconnected. The social and the natural sciences have indeed made great strides in their own specialties and each now has its own well-developed discourse. None the less, the problem is that they are talking past each other even though the sharing of data has become common practice among disciplines. The present chapter attempts to connect them under the rubric of the political-economy analysis.

This chapter discusses the relationship between environmental degradation and economic growth because fundamental aspects of the latter are implicated in the type and extent of urban contamination. At the core of this argument is the observation that the process of economic growth, and of globalization over the last decades, has much to do with the persistence of pollutants in urban skies. Reviewed here are the most relevant political and economic theoretical discourses that have criticized society for its treatment of the natural environment, and the most appropriate views for interpreting current air pollution in world cities are selected. Finally, a conceptual framework is developed to address the issue of continuous degradation of the environment and, following those theoretical guidelines, an interdisciplinary practical method to explore the accumulation of urban air pollution and related ill health is put forward.

Critical stances towards economic growth and its effects on the environment have emerged in sociology, economics and the political sciences. This interest has emphasized those aspects of modern society that are related to diverse aspects of environmental degradation. It has also promoted different solutions for conflicts between an increasingly industrialized and consumerist society, and nature as its sustenance base. The discussion of economic growth and environmental degradation has thus shifted from the continuing viability of the growth model – a consensus which, since the Industrial Revolution, has held that expanding production is a good thing; to neo-Malthusian 'limits to growth' concepts and the scarcity thesis of the early 1970s. It has moved to a critique of industrialization and modernization, to a search from the mid-1980s onwards for means to grow economically but within the ecological limits, through management and technological advance. Concepts and applications of sustainable development and ecological modernization best represent the latter. The social critique and suggestions that drive these theories are reformist. On the other hand, radical interpretations of economic growth and the environment have also been developed, and these draw on ecological economics and advanced Marxist sociological thought. A main objective of the current review is to identify the existence of a priori assumptions in the available theories, that is, the underlying thought which is normally ignored (Adam, 1994). Widely used theoretical views can be misleading and they naturally reproduce a divided or dual view of society, overlooking the precise links that this book attempts to trace. Useful interpretations, on the other hand, derive from acknowledging that biophysical and social factors intervene in the contamination of cities and recognizing that political and economic structures play key roles in environmental transformations.

Critical environmentalism and economic growth

As stated in Chapter 1, society and nature establish conflictive relationships due to the fundamental second contradiction of capitalism. However, different theories have not always viewed this relationship as antagonistic. This basic contradiction has been interpreted variously by either a pure descriptive and idealist, pragmatic and holistic, or a more comprehensive and realistic approach. Hence understanding of the theoretical basis that has dominated the subject has practical implications. Reformist thought, for example, has become the principle driving liberal economic policy-making in the developed world and has dominated the trend for integrating environmental considerations in existing economic sectors (for example, the Fifth Environmental Programme for the European Union). Three main planks of activity have been distinguished in the political and economic perspectives that link society and nature: symptomatic, reconciliatory and transformative approaches (see Table 3.1). The main theories that illustrate each approach are discussed according to their relation to the contradiction between nature and society, that is, through their consideration of the factors most responsible for environmental deterioration.

A first approach focuses on some characteristic aspects of economic growth while assuming that these represent the social origins of biophysical degradation. The approach that unifies the theories is

Table 3.1 A political-economy classification of growth and enviromental degradation theories

The approaches	*Leading theories*	*Origin of the problem*	*Typical authors*
Symptomatic	Limits to growth Industrialism	Natural scarcity Industrial society	Meadows; Ehrlich Porritt; Illich
Reconciliatory	Sustainable development	Management	WCED; Bartelmus; Elkins
	Ecological modernization	Insufficient and inappropriate technology and policy	Gouldson and Murphy; Spaarsgaren; Mol; Janicke; Jacobs
Transformative	Co-evolution	Disregard for connections of society, science and technology	Norgaard
	Materialism	Exploitation of nature through social relations	Dickens; O'Connor, J.; O'Connor, M.; Redclift; Gould, Schnaiberg

called here *symptomatic*. It postulates that the main underlying causes of environmental degradation lay either with the biophysical limits of nature – which reduce the possibility of producing more – or with industrialization of society, rather than with the deeper process of growth itself. One of the most common theories in the symptomatic approach describes factors that pose a fundamental threat to the 'carrying capacity' of ecosystems. It addresses concerns related to rapid population growth, widespread malnutrition, accelerating industrialization, depletion of non-renewable resources, and a deteriorating nature. Since the 1970s, the relationship between economic growth and the environmental crisis has been depicted as essentially a problem of scarcity on the global scale, concerned with too many consumers placing an unacceptable burden on a declining resource base. Limiting the growth of the economy is the well-known thesis of the Club of Rome, the most prominent representative of this standpoint. The other thesis with a symptomatic approach, industrialism, follows the central assumption that the development of industry and its impact on society are the central features of modern states and the main force behind the environmental crisis (for example, Badham, 1982; Porritt, 1984; Schumacher, 1974; Simonis, 1989; Singh, 1976). Industrialism theory understands the politics of the industrial age as a three-lane motorway, consisting of left, right and centre, with vehicles in different lanes, but all heading in the same direction of industrialization. 'The very direction is wrong', says Porritt, 'rather than the choice of any one lane in preference to the others. It is our perception that the motorway of industrialism inevitably leads to the abyss ...' (1984, p. 43). The same underlying thought is found in demodernizing authors, such as Illich (1975), Gorz (1983) and Bahro (1986), who consider that the industrial system is administered in an ever more centralized hierarchical way. Accordingly, the system is used as an organizational device that has become widespread but no longer adapted to the demands of human beings and of nature. They advocate a straightforward de-industrialization of society as the solution to all environmental and other problems. A basic assumption in the symptomatic interpretation of economic growth and environmental degradation is not to call into question the gratification of ever-growing human desires by way of a technologically mediated mastery of nature. This is a feature also characteristic of the next set of theories: while technology could not help realize the dream of infinite growth in a finite system, technology still represents the solution for the devil it created.

The symptomatic theories have raised important issues that link the natural environment with the workings of society. However, they do not go any deeper, such as into the general mechanism of the social system that has facilitated such occurrences as the growth of unregulated industry. The main limitations of the symptomatic approach are, first, that it reduces the cause of the conflict between environment and economic growth to a social and biological question. Second, it neglects the initial political and economic priorities that are instrumental in prompting leaders to choose measures that may cause excess use of raw material, malnutrition, degradation and accelerated industrialization. A belief characteristic of such theoretical views is that the real political reasons that people destroy the environment originate in a new social order based on technocrats and bureaucrats. This order of society is thought to have transcended the capitalist order and therefore capitalist societies have become industrial, post-industrial or post-capitalist. Indeed, the virtual inability of the symptomatic approach to appreciate the intricate relations between the environment and economic growth emerges because, by overlooking the economic forces behind industrialization, it does not transcend any political spectrum (Dobson, 1991, 1995). The symptomatic approach represents a general and shallow critique of the contradiction that gives rise to such persistent contamination in cities; yet it gives the illusion of a super-ideology within which communism and capitalism are inscribed. A different interpretation of the serious issues rightly raised by this approach is needed in order properly to understand and politically to challenge enduring air pollution in cities.

The underlying view of a *reconciliatory* approach unifies a second group of theories that also relates economic growth and the environment but from a different angle. It originates from the shift of emphasis in the 1980s – from the critical view of growth in all sectors of the economy to a debate on how to manage such growth in order to incorporate ecological and social considerations. The theories in this group find a middle course over the issue of economic growth and environmental protection. This approach avoids questioning the nature of economic growth but rather encourages it, suggesting improved and efficient patterns of production and consumption. Representative of this approach are sustainable development and ecological modernization.

The theory of sustainable development advocates the revitalization of global economic growth in order to reduce environmental degradation and alleviate poverty because existing patterns of development are simply not sustainable in the long term (Bartelmus, 1994; Stern et al., 1992). Sustainable development 'involves more than growth. It requires a change in the content of growth, to make it less material-and-energy intensive and more equitable in its impact' (WCED, 1987, p. 52). An important emphasis of sustainable development is on questions of even development and peace in developed and less developed countries, and not simply on economic growth as measured by per capita gross national product. Forests, soils and grazing lands were discovered as necessary input to long-term growth, but their availability can no longer be taken for granted. Sustainable development also implies that all negative environmental externalities such as pollution should either be internalized or reduced via technology, and there should be a non-declining stock of ecological and social capital (Elkins, 1993). Moreover, sustainable development ought to mean the creation of a society and an economy that can come to terms with the life-support limits of the planet, argue O'Riordan and Voisey (1997). The other reconciliatory perspective of economic growth and the environment is ecological modernization, a theory still under construction (Mol and Sonnenfeld, 2000). It rejects the notion of an incompatible opposition between capitalism – involving constant economic growth, technological development and the spread of consumerism through global marketization – and the goals of environmental conservation (Pepper, 1999). Ecological modernization stands for a major transformation, an ecological switch-over of the traditional industrialization process and consumption cycles to a direction that takes into account maintaining the sustenance base in the context of industrialized societies and uses more intelligent technologies (see Janicke, 1985; Mol, 1997; Simonis, 1989). The theory holds that a new phase of capitalist development is taking place, but with industrial society and industrialism as central to the development of modern society (Spaargaren and Mol, 1991). The view that environmental protection does not impede capital accumulation but has a potential to further accumulation clearly illustrates the reconciliatory approach (Cohen, 1997). Ecological modernization encourages the enhancement of lifestyles and domestic consumption (Spaarsgarten and Van Vliet, 2000). Social and institutional transformations have been, and still are, at the core of much current scholarship on ecological modernization (Elkington, 1987; Janicke and Weidmer, 1995; Pearce et al., 1989; Pearce et al., 1991). Ecological modernization promotes the application of stringent environmental policy as a positive influence on economic efficiency and technological innovation (Gouldson and Murphy, 1997). Similarly, rather than seeking a contradiction between society and nature, ecological modernization does not perceive economic development to be the source of environmental decline. It seeks to harness the forces of entrepreneurship for environmental gain.

The reconciliatory approach poses some difficulties as a prospective main theoretical tool for the analysis of urban pollution and economic growth in world cities. Its advocates consider that the environment can be a pool of resource inputs, even if these are scarce, rather than seeing it as a complex system that is transformed by development (Norgaard, 1994). Views that build upon a reconciliatory platform analyse the environment less as a commodity than does traditional economics which, drawing on market mechanisms, fails to allocate environmental services efficiently (Jacobs, 1994). Environmental systems are not divisible, almost never reach equilibrium positions, and changes are frequently irreversible (Norgaard, 1985). For reconciliatory theories, it is the externalities of growth (particularly, air and water pollution) and the efficient management of the environment, rather than scarcity, that provide the 'limits to growth'. However, the validity of efficiency in terms of ecological modernization has been challenged by the position that attempts to focus the social imagination on the revision of goals rather than on the revision of means for achieving efficient resource management (Sachs, 1997). For Sachs, the whole meaning of nature conservation changed with the concept of sustainable development. In a fresh and critical stance of sustainable development, he emphasizes that sustainability now means the efficient management of natural resources in order to optimize the yield of living resources, such as forests or fishstocks, by harvesting only so much as will not impair the rate of regeneration. For sustainable development, economic growth has to be seen as variable in its structure, while, on the other hand, nature has to be considered manageable.

The symptomatic and reconciliatory approaches of economic growth and environmental degradation have in truth advanced our understanding of the link between the natural environment and those aspects of the current economy that cause notorious environmental damage. Also, they appear to have the potential to allow environmental degradation and social intervention to become conceptually more closely linked than in scientific and reductionist views, as seen in the previous chapter (Redclift, 1992). However, the fact that they have done so, while failing to question the vested social interests that have protected and even encouraged polluting practices, means that these approaches lack a strong grip on what are the real sources of seemingly unstoppable urban contamination. The consideration of wider social processes such as globalization is essential for the assessment of such causes. We should not conceive of ecological modernization as a paradigmatic environmentally friendly shift in the economic production process without considering the articulation of local and sectoral shifts within the global context (Gandy, 1997). Neither can we ignore the fact that sustainable development as a desirable objective has served to confuse the costs that growth inflicts on the environment and, hence, on people. The two groups of theoretical approaches take for granted the central issue of economic growth as the only possible motor behind development, failing to recognize the direct, or indirect, damage that it may unleash on the environment. On this issue, Sachs says pointedly that the basis upon which the dilemma between the crisis of justice and the crisis of nature rests is the conventional notion of development (Sachs, 1997, p. 71).

A common feature of the reconciliatory and symptomatic approaches is that sustainability is discussed in purely economic terms. For example, O'Connor, J., (1994) has emphasized that this environmental economic tendency is present through the usual calculations of environmental cost and money capital, investment and consumption-related extraction of natural resources. For environmental economics, the current liberal social structures that have so enthusiastically empowered economic growth are seen as irrelevant to the deterioration of the environment; none the less, for them, these same structures certainly become very appropriate for overcoming the environmental problem. And from this economic point of view, sustainability economics must of necessity describe an expanding economy, something which contradicts in large part the steady-state economy that depends on the capacity of the environment, as suggested by the steady economy as postulated by Herman Daly (see Chapter 1). The interface between the natural and the social worlds becomes blurred in the symptomatic as well as in the reconciliatory approaches because they are founded on leading convictions that a single-line cumulative growth of scientific knowledge in history gives rise to a progressive mastery of nature through its application in technology, and the development of market forces as the means to ensure protection of the environment (Blowers, 1993). Sustainable development, none the less, can acquire a different, and very acceptable, connotation when applied to a perspective that is neither about economic excellence nor about biospherical stability, but about local livelihoods (Sachs, 1997). Creating high environmental standards, as ecological modernization does, can be a positive incentive to alleviate the contradiction between nature and society. Yet these measures are in fact supposed to promote new specialist industries and sectors, and to enhance the possibility of substantial employment growth. The reality is that, although urban air quality has improved since the 1950s, the ecological modernization that has taken place in the last decades has not been sufficient to reverse air pollution in cities.

The third and last approach consists of theories that consider that the causes of environmental degradation in modern societies lie in the macro-structural political and economic capitalist system of organization, historical evolutions in society, technology and science. We call them the *transformative* approach theories and they represent the most appropriate theoretical tool to understand economic growth and the environment in world cities. A transformative approach sees capitalism as the main problem underlying ecological questions – rather than as providing its solutions, as do the other theories. Two main perspectives offer explanatory approximations of the relation between economic growth and environmental degradation that are most useful: co-evolutionism and materialism.

The transformative approach sees that scarcity as a natural biophysical limitation, and industrialization as one of the main sources of environmental degradation, must be located in historical and social contexts (Dickens, 1992). The point here is that, as Harvey's early writings stressed (1974), scarcity in developed societies is not a product of certain levels of taste and consumption, but rather deliberately created and manipulated by producers. Scarcity is necessary to the survival of the capitalist mode of production. Rather than scarcity as the cause of the environmental crisis, argues Redclift (1987), the costs of environmental degradation and its distribution are such that continued growth becomes unacceptable. As a consequence, long before it becomes biophysically impossible to grow economically because of insufficient resources, it becomes socially undesirable to do so. Economic growth is, in fact, the reality that makes human choice less and less possible under conditions of scarcity (Redclift, 1987).

The main theoretical perspective that is built upon a transformative approach is materialism, which is best represented by authors who have drawn on, and significantly enriched, original Marxist interpretations of the relation between environment and society. The capitalist mode of production has a central role in both the generation of this environmental degradation and the impossibilities of overcoming environmental devastation (Gould et al., 1996; Pepper, 1999). Much of traditional political-economy theory – at least until the 1970s or 1980s – has tended to underrate the importance of social variability within social (capitalist) structures. It has been necessary to incorporate new concepts and understandings concerning the relation between nature and modern society. Dickens warns that 'turning from functionalism to Marxism did not mean that the analogies and dualism between nature and society were finally dropped' (1992, p. 49). Traditional theory has limited an appropriate appreciation of the varying ways in which real capitalist societies have accommodated the self-destructive forces and contradictions of capitalism (Dickens et al., 1985; Dickens, 1997). Such narrowness meant that only later was the second contradiction of capital, that between society and nature, brought to light. As stated in Chapter 1, it is this fundamental and general framework that explains the current deteriorated state of the environment under conditions of economic growth.

In the main, the materialist political-economy tradition has been concerned with explaining the context in which the approppriation of nature takes place. Schnaiberg and Gould (1994) and Gould et al. (1996) centre the analysis on the 'treadmill of production'. The treadmill concept implies that the process of production tends to result in environmental degradation through withdrawals from nature, and additions, that is pollution. The treadmill of production concept holds that modern capitalism and the modern state exhibit a fundamental logic of promoting economic growth and private capital accumulation and that the self-reproducing nature of these processes causes them to assume the character of a 'treadmill'. But clearly, stresses Redclift (1984), 'what could not have been predicted in the lifetime of Marx or Engels was that capitalism would pose such a threat to natural resources that the very existence of development would be called into question' (p. 6). Even if well managed, declares this view, it is a fallacy to believe that any level of economic growth is possible without a corresponding degradation of the environment, however limited (Redclift, 1987). The contradiction between society and nature wrecks the environment through flows of trade, industrialization, investment and aid that encourage the exploitation and waste of resources, pollution and destruction of the ecosystem.

Co-evolutionism offers appropriate insights into the adverse effects of industrialization and stresses the reciprocal effect of joint social, economic and technological change. Crucially, it breaks with the supremacy of the paradigm of the nature/society divide, an issue not particularly relevant for previous approaches. The era of hydrocarbons drove a wedge between the earlier co-evolution of social and ecological systems when capturing the energy of the sun through ecosystem management became less and less important as Western science facilitated the capture of fossil energy (Norgaard, 1985). Economic systems co-evolved with the expanding number of technologies for using hydrocarbons and only later adopted institutions to correct the detrimental transformations this co-evolution entailed for ecosystems and ultimately for people. Co-evolution explains, for example, climate change, which unlike the environmental concerns of resource scarcity in the 1970s, was prompted by apparently profligate global energy consumption (Redclift, 1996). The co-evolution theory explains that hydrocarbons freed societies from immediate environmental constraints but not from ultimate environmental constraints, that is, from the limits of the hydrocarbons themselves and of the atmosphere and oceans to absorb carbon dioxide and other greenhouse gases associated with fossil-fuel economies (Norgaard, 1994). The co-evolutionary approach recognizes that societies in the twentieth century can be characterized as social and economic systems whose growth is co-evolutionary, based on stock resources and the neglect of environmental systems (Redclift and Woodgate, 1994). Norgaard emphasizes this interactive synthesis and equilibrating operations; but, significantly, he adds that at times, there are unexpected ecological results of both natural and social mechanisms of change (Norgaard, 1997).

In summary, the three identified critical political-economy approaches to economic growth and the environment contributed material for assessing the causes of persistent contamination in modern cities. Opting for a symptomatic approach would mean putting responsibility for persistent urban contamination on the limited capacity of the atmosphere to absorb industrial and vehicle traffic emissions, rather than on the interaction between regulations and economic priorities of governments, corporations and other interests. Adopting a reconciliatory approach would mean blaming inefficient management and ways of production for excessive pollution in cities. A transformative approach focuses on the social forces that act to continually modify the environment, and identifies the priorities of society that affect what and how to produce, distribute and use.

Reorganization of global economy and urban environment

In discussing current economic growth and the natural urban environment, the globalization of the economy must be incorporated because a new world economic order has emerged since the 1970s, with evident implications both for the global and, significantly, the local environment. A critical appreciation of the process of globalization (Chapter 1) is thus articulated to the theoretical transformative approach to explain current pollution in cities. Globalization represents a major and far-reaching strategy for expanding economic growth. It affects the geographic, economic, cultural and social dimensions of current society and encompasses changes such as relocation of industrial plants; a burgeoning urbanization that consumes green belts and appropriates built urban land as if these were empty sites; and internationalization of financial markets. The unprecedented reorganization of the world economy has generated visible changes in built environments. Whereas there is now a lower level of urban pollution in comparison to the mid-twentieth century, this decrease in environmental degradation has been offset by the continued growth in output. Industrial processes, international trade and other activities connected to increased output are not unrelated to conspicuous and also less visible degradation of the natural environment. In this constellation of growth and globalization, world cities occupy preponderant functional economic roles, a point that is further illustrated in Chapter 4. At the same time, world cities are also condemned to accelerated local environmental degradation, as this book shows for Houston in particular.

Importantly, a justification for recognizing the strength of globalization in current structures only derives from the fact that today globalization represents a major factor for change – that is, a mechanism comparable to that of alcohol which causes drunkenness in people (see Chapter 2). Globalization adds a further dimension to the basic contradiction between nature and society. This component is geographic and also visible in world cities, and is important for observations of actual environmental and health transformations in one city. Following from the points raised in Chapter 1, economic intensification has been led by multinational companies, banks and state governments (Thrift, 1988). It has engendered a deep international division of labour, new patterns of consumption and needs, the incorporation of new spheres of production into the global market, relocation of industrial and energy plants in the developing world, and a greater centralization of world finances. The spread of information technology, as a major dimension of globalization, has been an essential component for the 'success' of world-financial operations (Yearley, 1996, p. 4), the heart of the productive capacity of economies, and the military might of states (Castells, 1998, p. 320). All these are aspects of globalization that have affected, directly or indirectly, the quality of the environment not of one, but of many cities worldwide.

It has been argued that some elements of a more globalized economy may initially seem benign (or even positive) for the environment, although the opposite may be true when viewed in a more dynamic way and in the longer term (OECD, 1997). From this viewpoint, one may conclude that the analysis of the environmental consequences of globalization must be carried out over a longer time frame than is required for many economic issues. Some considerations might be that the rate of globalization is not constant over time and, similarly, the rate of environmental change due to globalization may not be constant. However, this book argues that the current deteriorated state of urban environments already offers a sufficient testimony to the adverse consequences of globalization. In fact, the process of environmental protection might well be delayed if longer-term environmental assessment is implemented. Such longer terms are allegedly needed to obtain further scientific information to reduce risks. But realistically, these will just enable undisturbed growth. Increasing air contamination in world cities like Houston, London, Philadelphia and New York is a realistic indicator of the lasting strategies of economic growth as well as of the local outcome of globalization practices. Important foundations promoting environmental protection are assessment of current and past health risks associated with urban environmental hazards as evidenced by the epidemiological data, the analysis of the role of social structures, and the cognitive need to link expert and lay knowledge. The book positions the immediate issue of urban air pollution and health as integral to world cities and argues that a transformative political-economy approach and biological framework are needed within a spatial perspective that looks at interaction of geographical levels with social structures.

An approach to urban pollution

The social and biophysical dimensions of life in world cities are connected through numerous and significant interfaces. Yet society as much as nature displays independent realities which coexist in cities; therefore the input of the biophysical approach in terms of environmental science is needed. As seen in Chapter 2, there actually are entities and causal processes that cannot be contained within social theory (Dickens, 1997). On the other hand, the contention is that economic growth and environmental pollution are intricately linked (Cherni, 1993b). By separating these two components, a serious risk arises of misjudging causes and the possible solutions of the familiar configuration of pollution in cities. It is necessary to avoid a scenario in which the phenomena of increasing environmental degradation and ill health continue to be treated as if they were divorced from the dominant type of economic activity. Also the main political priority of government, and our alienated and divisive relations with nature, should be recognized for their place in the analysis. The idea of a 'weak biological determinism', to use Benton's terminology (1989, 1994), has therefore been adopted for analysing economic growth, globalization and environmental pollution in cities. Weak biological determinism commits the analysis of social structures to a natural sciences approach but not necessarily to a reductionist one. It recognizes the relevance of evolutionary theory, physiology, epidemiology, genetics and, especially, ecology, as disciplines whose insights and findings are pertinent to our understanding of the natural environment, ourselves and society. Remarkable works such as those by Rose et al. (1984), Benton and Redclift (1994), and Lewontin and Levins (1996, 1997) have contributed to the critical analysis of the relation between this biological perspective and society.

Human health is one aspect of society which researchers have typically associated with biophysical conditions, public institutions, indicators of poverty, or with individual behaviour. Research on pollution and health has produced very useful information but piecemeal explanations, as has been discussed in Chapter 2. Showing that airways become obstructed due to the effect of airborne particles is crucial, yet it tells us nothing about the political and economic mechanisms and the social commitments that permit certain levels of air particles in cities. Therefore, the approach of ill health and the environment must be rectified in order to explain the problem of long-lasting urban air contamination, with the objective of improving the well-being of residents in urban areas. Applying a transformative approach and an appreciation of the process of globalization, one can explain the biological and chemical changes occurring in the local environment and the consequences for residents' health. For example, commodity production itself will affect health in a variety of ways: the biophysical repercussions of shift-work, de-skilling, overtime or the use of dangerous chemicals (Doyal, 1987). Yet, the biophysical effects of the production process extend beyond the work place because health-damaging materials are emitted into the environment in the form of industrial externalities. Despite adverse health effects, commodities continue to be made, and fossil energy to be produced and consumed, simply because they are profitable and have become entrenched and essential to our lifestyle. Navarro (1986) has suggested that many of our health problems exist because of and are reproduced by the process of capital production, which expands and affects the world of consumption as well, becoming the problem of everyday life. The continuous deterioration of nature in the urban environments that results from pollution of various kinds is often the by-product of industrialization and of urbanization. The two respond to development policies promoted by governments as part of the institution of economic growth and globalization. The actual limits of relentless economic growth are not only resource scarcity, deficient management of nature, lack of stronger policy, as symptomatic and reconciliatory theories argue. It is important to think of the limits in terms of those aspects of society that promote unending production to satisfy undefined goals. These activities can pose numerous risks in human and environmental terms.

Many causes of the environmental crisis in cities are structural, with roots in political institutions and relations of production; hence anything other than a political and economic treatment of the problem of urban air pollution and human health lacks credibility. The theoretical argument of the second contradiction of capital explains the position of conflict between nature and society. Moreover, co-evolutionism helps our understanding of changes in nature by taking into consideration the historical context. An acknowledgement of this type contributes to a resolution of the issue of ahistoricism that is often inplicit in the scientific approaches that have addressed air pollution and disease. As society has increasingly made an impact on the biological and biophysical worlds, so it becomes increasingly necessary to develop theories and understandings that cut across these different spheres.

The last point is that the concrete is not something which is reducible to the empirical (Sayer, 1992). Our concrete objects of study, that is, air pollution and health, are concrete not simply because they exist, but because they are a combination of many forces or processes that are activated over space. The abstract logic of the process of economic growth (for example, the promotion of free market competition and the search for profit, the neo-liberal state and deregulation), articulates with material spaces and people to generate unique concrete social spatial patterns of production, residence, and varied forms and degrees of environmental degradation. These specific population and environmental urban patterns need to be understood before the presence of social and political processes at work can be identified. In this sense, the real is more than the concrete, for embedded within it also lie crucial social processes which have enabled biophysical change to occur. This book looks into the real and dynamic world of air pollution and ill health in order to challenge ahistorical and detrimental misconceptions that have led to the misguided belief that the most valid considerations are the biophysical.

A polluted city like many others

Houston provides a useful case study as a contingent urban setting within which the relationship between health and pollution may be examined. Houston is considered as a microcosm of both economic growth and environmental degradation. It epitomizes the capitalist heaven and the air pollution haven recognized in the literature (Blowers, 1993). This 'paradise/hell' contradiction is shared by many world cities today, like London and Tokyo. This relationship is also typical of cities like New Delhi, Mexico City and Bangkok in the Third World (Hardoy and Satterthwaite, 1987).

Although there are a number of noteworthy points about Houston's successful growth in particular, the city is certainly not considerably different from other major US cities such as Los Angeles, Dallas or Detroit. As in other American and European cities, urban growth and air contamination illustrate dramatic changes occurring in most metropolitan areas in the world over the last decades as a result of unprecedented developments in the local and global economy. Yet, focusing our attention on one selected city is important for empirical as well as theoretical reasons. Indeed, different structural processes can produce similar empirical results, just as the same processes at the structural level can produce different practical results. The quantity and quality of concentrations of air pollution in the real world are important because they can make a difference to the type and extent of health damage; and because they indicate technological stages and social commitments. It is thus plain that neither pure empirical nor theoretical studies alone can explain the relationship between economic development, air pollution and ill health. Sayer (1992) emphasizes that

> Explanations of concrete phenomena which abstract from spatial or other forms must be regarded as being significantly incomplete. Yet few social scientists even recognize the problem, and this despite the fact that variations in form are a major factor in the failure of causal mechanisms to produce empirical regularities. (p. 247)

A useful and genuine picture of reality requires quantification as well as qualification of the main events of air pollution and ill health. A central source of this information is the local public; indeed, the incorporation of lay knowledge is essential to avoid reproducing the dualism of the sciences (Chapter 1). At a procedural level, political-economy theory of economic growth and the environment, the conceptions of interaction between social and biophysical dimensions, and assessment of the concrete events can all be illuminated by a case study. Residents' reports enable the researcher to point out whether there is a need to address loopholes in the legislation, and to assess what is wrong with the environment and its people. Therefore the underlying commitments of governments, the political strength of economic corporations, the pure dedication of the research community and its relative position in society, are all reflected in the state of the environment.

Applying knowledge from a number of relevant disciplines to describe and explain the persistent air pollution at the beginning of the twenty-first century in world cities requires a cohesive and critical approach. Figure 3.1 shows a general quadrate conceptual model

Figure 3.1 Interdisciplinary framework for environmental degradation and risk

that consists of theoretical, contextual, participatory and concrete dimensions. It is dynamic and indicates the interactions among the sectors, but also shows the prevalence of the political and economic analysis for explanatory purposes. The model indicates the interdisciplinary nature of the approach. It also represents a political reconceptualization that contributes to the resolution of the enigma of persistent contamination in cities. The model attempts to capture the uniqueness of each city case as well as the generalization of social and biophysical processes. Therefore, the model can be replicated elsewhere.

Table 3.2 gives a practical guide to research that connects various facets of the problem. Both quantitative and qualitative analysis are necessary in this type of interdisciplinary and political enquiry. Quantitative information enables a more precise description of the state of the environment, of public and child health citywide and in study areas. Qualitative knowledge identifies the views and accounts of residents and key people, assesses social structures and processes into which individuals are locked, and incorporates the alterable mechanisms of nature that intervene here. Both types of research are important but they fulfil different functions, the one primarily informative, the other primarily explanatory (Gregory, 1986; Sayer

Table 3.2 A multi-faceted model for quantitative and qualitative research into air pollution and ill health

Research areas	*Population – Residents* (1)	*Socioeconomic structures* (2)	*Globalization – context and processes* (3)	*Biophysical features* (4)
Indicators	Financial	Local economy	Sectoral economy	Pollution
	Demographic	Public health	Competitiveness	Local degradation
	Geographic	Environmental	Global environment	Climate conditions
	Hazardous exposure	Regulation	International role of city	Urban landscape
		Power and protection		

and Morgan, 1985). Yet, to be useful, they must complement each other and the 'multi-faceted model' represents a practical guide for this type of research. Paradoxically, it is exactly the process of deconstructing the taken-for-granted assumptions surrounding the question of environment and society, and of breaking them down into constitutive parts but incorporating them in a dynamic structure which, in turn, makes explanation possible. This process is necessary to make the urban environmental problem comprehensible, and to clarify social responsibility.

Conclusion

This chapter has captured significant aspects of the multi-faceted configuration economy and environment, adjusted the separation between the sciences, assimilated both the utility and the criticism of basic scientific theories and methods, interpreted social and other contingent events, and provided explanations for historical and spatial patterns of air pollution and of child ill health. Appropriate scientific explanations can certainly be instrumental in challenging the persistent problem of urban air pollution and ill health in contemporary society. Yet, to achieve such explanations requires an interdisciplinary perspective that is firmly grounded in a critical social approach that recognizes the power of political and economic institutions to affect biophysical conditions. The interrelations carry most weight in explanatory terms and these are a main focus of this book. Air pollution, like declining water quality and vulnerable ecosystems, must be seen within the framework of the limits placed on economic growth, which are increasingly not only those of resource scarcity but of plenty, particularly in the form of hidden externalities (Redclift, 1996). These threaten to undermine the very economic systems from which they derive. The chapter has discussed the available analytical perspectives and justified appropriate theoretical and methodological designs. The second contradiction of capitalism is at the very basis of the relationship of air pollution in cities. The conceptual framework draws, however, on an elaborated model that considers four dimensions of sustainability. The theoretical framework is then projected into a practical model for research of local transformations of the quality of the air and public health. This relation is assessed by analysing and connecting development of regional industries, settlement of residential areas, and levels of urban pollution in regional and global perspectives. The analytical model follows the logic of inclusion. It encompasses the participating chemical, ecological and biological mechanisms, current and historical social structures at local and global levels, empirical information of the events that occur in a city; and knowledge from local people.

Our fundamental task of quantitative analysis is to investigate the trends and dimensions of the events, and contextual and qualitative knowledge reveals relations of causality. The two research approaches complement each other in describing and explaining the phenomena under investigation. This approach is strongly shaped by the methodological premises of critical realism, and the multi-faceted model consolidates the different aspects of the problem. The next chapter moves on to describe the processes that have taken place in a current urban context which have made it a world city.

4
Globalization and Local Change

Introduction

The theoretical analysis of Chapter 3 mentioned nature as something that is crucially affected both by the economy and by policy. But what the 'something' is, why and how it is being ruined, and how it relates to human beings, have hitherto remained inadequately discussed. This chapter describes the processes that affect the environment and how these make the identified contradiction between nature and society more explicit. We discuss the transformation of Houston into a world city as part of a general trend taking place in other developed countries. The focus in Houston has been on the rise of the profitable, but highly contaminating, oil industry in the region. The petrochemical and other related businesses have played crucial roles in the growth of the city and its involvement in the global energy business. They have ensured that Houston has fulfilled specific functions for the international economy, irrespective of any ecological or human cost. The extent to which residents have benefited or, conversely, suffered from local and global directions in the economy, is difficult to assess. This chapter evaluates what for decades has been accepted as causing only benefits – economic growth.

It becomes difficult to understand the actual forces that have underlain environmental degradation, and air pollution in particular, in large cities, usually because these are complex. Visible and less visible biophysical and social mechanisms interact along a continuum that affects the environment. Economic, political, chemical and biological factors are frequently found acting together and, often, mutually reinforcing each other. It is, however, the biophysical aspect of the interaction that is most often scrutinized when it comes to urban air contamination and health. Making reference to a particularly large and economically competitive city, this chapter looks into the onset of environmental contamination by discussing the social processes that brought significant changes to the quality of the physical surroundings and of the air. The purpose of this chapter is not, therefore, to give an absolute and final explanation of the process of environmental degradation, but to highlight crucial political and economic processes which have hitherto been largely unaccounted for.

This chapter broaches the idea that the time and space intervening between the making of an urban capitalist heaven, and its transformation into an urban environmental hell, can be, in fact, negligible. In many world cities, these two phenomena have grown in tandem, as the case of Houston illustrates. Some of the aspects of economic growth and globalization may be perfectly obvious in world cities. For example, the continuous demolition of old-style buildings leaves empty land that is used for the new cycle of capital investment – often for constructing modern offices to house international corporations. Another visible example is the change from previously nationally managed utility companies like electricity to international ownership. However, many other effects, such as changes in the quality of the air, remain mostly unrelated with the same process. A commitment to liberal business practice has barely been associated with anything other than economic achievements, job opportunities and creation, wide access to communication and transport, improved international competitiveness, and overall urban splendour. It is within this line of thought that Houston has gained a reputation for outstanding success in the international and global economy, while the fact that the quality of the air has remained dangerously contaminated has been ignored. The severity of air, as well as water, pollution, and also the socially unequal distribution of wealth resulting from this type of 'successful' economic growth have, in fact, been hidden behind the city's economic prosperity and high material standards of local living. Indeed, while two of the five wealthiest suburban areas in the US are situated in Houston, huge areas of poverty, large minority ghettos, severe environmental pollution and remarkable ill health also characterize the city.

The urban natural environment in Houston is an aspect of the city that has been adversely affected by a century of uncontrolled economic activity and pro-growth policy. This chapter argues that the social processes that have brought remarkable growth to the region are the same ones that have caused notorious pollution. A transformative perspective advances the understanding that society as a system that employs particular technological and scientific designs is intricately linked to recognized and also less recognized environmental damage. It is argued, therefore, that, conceptually, two processes take place: the interdependence between the social and biophysical systems as explained above, and the development of economic enterprises and political organizations over both time and geographical space. Along this course of interdependence and development, a number of contradictions have become evident. One contradiction is that between the affluence that has been achieved due to unquestionable competitive economic growth, and its opposite, the contamination and dereliction springing from the same process. The other contradiction refers to two contrasting realities in the city, one being shaped by an auspicious and strong linkage to the global economy, the other showing local social inequality, uneven distribution of industrial pollution, unequal access to outstanding medical facilities, and low public health achievement rates. The chapter explains the coexistence of a capitalist heaven and a pollution hell from the political-economy perspective of political ecologism and globalization. It highlights striking urban contrasts and perplexing environmental contamination.

Globalization of the energy economy in Houston

A shifting international division of labour has emerged since the late 1960s and 1970s, but it is not, in fact, new. For at least two centuries, business people and industrialists have expanded operations across state boundaries to exploit raw materials, labour, production sites and overseas markets (Feagin and Smith, 1987; Thrift, 1988). In recent decades, however, a new international division of labour has increasingly involved transnational firms investing heavily in many countries and trading goods and services with one another, or transnational firms' subsidiaries trading within the globally extended framework of one large corporation. Major externalities, such as air pollution at the local level, can happen because international corporations, for the most part, calculate profit and loss at the firm, and not the societal, level. Environmental externalities are allowed to occur because governments have fostered economic policies with the purpose of achieving global economic status, rather than reducing or stabilizing toxic emissions. Global cities constitute vivid expressions of this social process. Geographical space has been the linking factor in this relation simply because the political and economic processes that permit growth develop over space; therefore, space becomes socially constructed (Chapter 2). Growth and the city of Houston have been put together through the international, regional and local economies. This characteristic, which is becoming an increasingly common feature of major cities such as London and New York, needs to be incorporated when the analysis of local pollution and ill health is carried out. It is an incipient characteristic of cities which can be linked to Feagin's (1988) interpretation of contemporary cities. He points out that places like Houston are not islands unto themselves; rather they are greatly affected by capital investment flows within the regional, national and international contexts. Massey (1995) emphasizes this point when she argues that local differences and uniqueness are defined in terms of the interdependence with international relations. It is therefore crucial to address globalization, but also local events, in conceptual and empirical terms.

The global context of local growth

The beneficial configuration of oil and gas, sulphur and water, and available labour in the Houston area certainly contributed to the development of the petrochemical industry in the South (Child Hill and Feagin, 1987; Feagin, 1988; Thomas and Murray, 1991).[1] However, natural advantages alone could not spur growth in the region. Beyond the recognition that the presence of the natural quartet of oil and gas, sulphur and water, all in one place, was essential to the development of the industry, the economic and political forces acting on behalf of the national government and international firms have also been crucial to the way the region grew. As a matter of fact, politicians and industrialists could have exploited the biophysical advantages in the Houston region for a different purpose and also in a different manner – or not exploited them at all, which would have spared the environment and local residents plenty of unpleasantness.

Two mainstream theories have frequently been employed to explain growth in Houston: those of development, called *convergence* theories; and those of critical power conflict, termed *uneven development*. Convergence theories have emphasized that Sunbelt cities such as Houston are 'catching up' economically with northern cities, for example, New York, which developed earlier, and that this convergence is part of a tendency towards equilibrium in US society. Sunbelt cities like Houston are seen as late-comers, their rise the result of decentralization to the hinterland, and substantial in terms of technological changes in transport and communication, with prosperity accomplished in the absence of state intervention (Hawley, 1981; Kasarda, 1980; Rostow, 1977; Williamson, 1965). The uneven development perspective, on the other hand, has stressed uneven development as a normal aspect of cities embedded in the capitalist system (for example, Smith, 1984). Perry and Watkins (1977) argue that Sunbelt cities have prospered by attracting leading industries, including electronics, defence and oil. Northern, as opposed to southern, cities have failed to capture these industries because of the commitment of investors to established industries in the North. The theory claims that development in one city is achieved at the expense of cities in other regions and that government plays a role in how intervention makes an impact (Hill, 1977; Perry and Watkins, 1977).

The two theories are, however, insufficient to explain growth and concomitant environmental degradation. The geographical scope of their analysis expresses a problem. Emphasis on the context of globalization, as stated in Chapter 3, has been lost. This framework is important because 'capitalist development transforms nature and the environment within a logic which needs to be understood in global terms' (Redclift, 1987, p. 47). The transnational corporation has been the major locus of transnational economic practices – the foundation on which service-related firms and a huge array of small to middle-sized commercial and industrial corporations have been grounded – and the transnational capitalist class is the major locus of transnational political practices (Sklair, 1994).[2] Control of most of the factors that influenced the growth of the region's refining and oil industry was generally vested in the distant headquarters of nationally and internationally active corporations.[3] With the absence of top oil firm headquarters but a high concentration of major divisions of top oil firms, the role of Houston, like that of Detroit, in the international division of labour has been that of a 'divisional command city' (Feagin, 1985; Feagin and Smith, 1987, pp. 6–8). Within the context of the international economy, the role Houston has fulfilled is that of an oil-specialist international city:

> The world of modern capitalism is both a world-wide net of corporations and a global network of cities ... But most cities are not at the world command level; indeed, different cities occupy a variety of niches in the capitalist world economy.
>
> (Feagin and Smith, 1987, p. 3)

Indeed, the long-distance intercity relationship was not one of filtering growth down an urban hierarchy but rather of moving corporate investment – often drawn from outside both cities – linked by a manufacturing need for processed raw materials (Feagin, 1985). The internationalization of the world city developed with ties outside the US economy when the major oil companies began to develop international operations in the 1920s and 1930s. The Gulf Coast oil industry moved quickly through the stage of competitive-industrial local capitalism to that of oligopoly capitalism dominated by major companies (Feagin, 1985; Thomas and Murray, 1991). By the late 1920s, 70 per cent of Texas production was in the hands of 20 companies, although there were 14 000 oil companies throughout the US (Williamson et al., 1963). In the post-war period, the growth of Houston was facilitated dramatically by the federal government, which had become a primary source of capital for, particularly, oil-related development previously linked to the war (Shelton et al., 1989).[4] This indicates that the spatial form and environmental characteristics of the city of Houston are not some easy and inevitable translation of deep-lying economic structural forces, since specific economic and state forms do not develop inevitably out of structural necessity. They develop in a contingent manner as the result of the conscious actions taken by members of various classes, acting singly or in concert, under particular historical and structural circumstances. Spatial forms are contingent on the dialectical articulation between action and structures and 'structural changes are processed through human actions taken under historical conditions' (Gottdiener, 1985, p. 199).

In fact, as a result of shifts in the world oil market in the 1960s and 1970s, scattered company operations around the US were closed and consolidated in larger offices in a few key cities, New York, Los Angeles, Detroit and also in Houston (Feagin, 1985). Of the world's 35 largest oil companies in the 1960s, 34 had located major office and plant facilities in the Houston area.[5] Since the formation of OPEC in 1973, the city replaced its role of direct involvement in oil processing with a much greater emphasis on oil-related, often export-oriented, services, technology, financial policy and control functions (King, 1991). Hundreds of national and international geological firms, drilling contractors, supply companies and law firms, together with 400 major oil and gas companies, were based in the city (Child Hill and Feagin, 1987). Substantial international trade passing through the Port of Houston linked Houston and cities in Latin America, the Middle East, and the Far East. The city's economy, with its worldwide network of production, exchange, finance and corporate services, came to be based on transnational practices (Feagin, 1988; King, 1991). Houston housed 11 top company headquarters in 1984, representing the ninth largest concentration of major multinational corporations in the world (New York housed 59 top firms; London 37; Pittsburgh and Hamburg, 10 each; Toronto 7).[6]

The federal government and local entrepreneurs played a crucial and ongoing facilitating role in the rise of the Gulf Coast city after 1902, a role that has so far continued unchanged into the new century. Essential infrastructure for the development of the oil industry, such as port facilities, the Ship Channel and city roads, were improved with federal funds. Governmental subsidies were secured by local bankers, real-estate investors, top governmental officials, and other leaders – the so called 'growth coalition' – who lobbied for government monies, contradicting, in fact, the growth coalition's professed, but not practised, philosophy of free enterprise and no government intervention to preserve the business climate in the global city of Houston (Feagin, 1988).[7] In this sense, the active role of the 'growth coalition' was very important for enhancing the world competitiveness of the city to attract further business. Massive defence spending on the Pentagon's production system and the military–industrial complex spawned new industries in the city such as high-tech complexes and suburban office–commercial parks. 'Houston's greater ability to garner largesse from the federal government is a windfall passed out by the Pentagon' (Child Hill and Feagin, 1987, p. 174). In summary, 'cheaper production costs (weaker unions, lower wages), weaker biophysical and structural barriers to new development (no aging industrial foundation) and tremendous federal expenditures on infrastructure facilities (highways) and high-technology defence industries' (Feagin and Smith, 1987, p. 168), contributed to encourage Houston's growth and to make the area the supplier of one more cheap commodity for the national and international market: oil was added to cattle, cotton and timber. Clearly, environmental protection could not become a priority in a city where the means to achieve growth and profit were unrestricted and uncontrolled, where the national government actively encouraged these aspirations, and, significantly, where the principal forces of growth were tightly linked to a global system of increasing energy demand.

The importance of the growing oil industry

Houston has played a leading role in the development of the US energy sector since the 1960s until the present, and has been a notoriously attractive location for the global reorganization of the oil and gas industry. In the 1990s, about one quarter of the US's oil refining capacity, one quarter of the oil–gas transmission companies, and one half of all manufactured petrochemicals made in the US, were located in the Houston–Gulf Coast area. The general trade passing through the Port of Houston reflected extensive commercial activity in the area. The port ranked first in the US in terms of foreign tonnage and second in total tonnage and it was the tenth largest port in the world (Port of Houston Authority, 1995). By the mid-twentieth century, the Port of Houston was already one of the world's largest industrial concentrations. Figure 4.1 illustrates the rising trend since the 1960s.

Historically, population growth rates between 1860 and 1990 in the Metropolitan Statistical Area (MSA) – comprising the surroundings counties of Montgomery, Fort Bend, Liberty and Waller – have been above the US's (Greater Houston Partnership, 1992). In 1990, Houston was the fourth most populous city in the US (2.8 m) and the largest in the South and the Southwest. Annual population

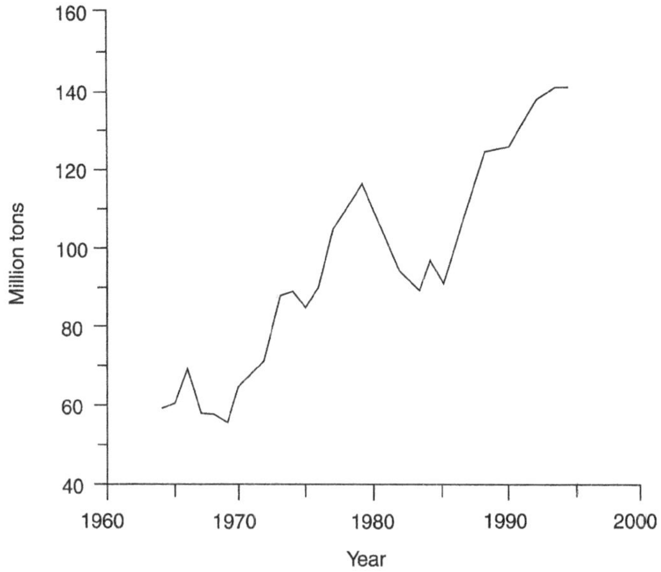

Figure 4.1 Cargo tonnage in the Port of Houston, 1964–94
Source: Data from Port of Houston, US Corps of Engineers, MR 14/3/95

growth was remarkably constant, at about 3 per cent per year. Population growth per decade in the period between 1850 and 1980 was a striking average increase of nearly 50 per cent for the metropolitan area (Feagin, 1988; Thomas and Murray, 1991).[8]

Houston's economic base was affected as much by international as by national events – the 1973/74 OPEC price increase, for example, brought rises in profits for oil companies, boosting the Houston base. Conversely, a major recession affected Houston in 1982–87 due to the decline in oil prices in the early 1980s, the construction of numerous large-capacity refineries in the Third World, including the Middle East and Asia, and the problems of declining oil production, oil glut and slowing of demand for oil in the US (Thomas and Murray, 1991).[9] However, the economic recession in the 1980s did not result in a movement away from national and global oil and natural gas markets. Houston continued to provide raw materials, specialized services and markets for the global capitalist economy which have been qualified as essential in the new international division of labour (Feagin, 1985).

Crude oil operations increased consistently in the 35 years of operation between 1941 and 1976. The total refining capacity of the Houston area rose more than in any of the four competing areas on the Gulf Coast (Houston, Beamont–Port Arthur, Corpus Christi, and Louisiana Gulf) (see Table 4.1). Houston enjoyed a long boom period in the aftermath of the Second World War based on the rising global demand for oil products; construction companies, truck, pipeline and shipping companies grew around Houston oil; most of the US's largest oil companies located administrative, research and production facilities in this metropolitan area; and there has been collaboration since the 1950s between Houston oil companies and oil fields, from Malaysia and Saudi Arabia to the North Sea. Most large oil fields opened within the global economy brought new growth to Houston.

For Houston, the contamination of the natural environment has been concomitant with filling this privileged position in the international economy. Local business people, international corporations, and also the national government have clearly ignored this process of severe air contamination. Since the city became the

Table 4.1 Operating crude oil capacity of Houston on the Gulf Coast, 1941–76

Year	*Houston* production*	*Houston per cent of Gulf production*	*Beaumont – Port Arthur per cent of Gulf production*	*Corpus Christi per cent of Gulf production*	*Lousiana Gulf per cent of Gulf production*
1941	472 800	40.5	38.6	9.1	11.8
1947	539 300	34.4	37.1	6.0	22.5
1951	740 200	36.2	35.2	4.6	24.0
1956	925 600	34.3	31.3	8.3	26.1
1961	946 800	32.3	32.7	8.6	26.5
1966	1 085 350	33.8	32.3	7.6	26.3
1971	1 453 500	33.8	28.6	7.9	29.6
1976	1 941 000	35.9	22.8	8.7	32.5

*42-gallon barrels per day

Source: Adapted from US Bureau of Mines, Petroleum Refining Information Circulars in Pratt (1980), p. 96.

centre of a national and international economy, 'local concerns mattered less and less to the corporate leaders who ordered and controlled the direction of industrialization and who focused on national and international markets, but not on local matters' (Fisher, 1994, p. 3).

The energy capital of the world

Houston had long played a prominent role in the oil business. Already in the 1970s it had emerged as the unofficial 'energy capital' of the nation, which was the logical culmination of over three-quarters of a century of oil-led development (Pratt, 1980, p. 5). None the less, at a time of oil abundance in the world economy at relatively stable low prices, the move of major international oil subsidiaries to Houston and the buttressing of existing operations there during the 1960s and early 1970s was a new element, part of a series of significant changes in the organization of the global economy (Cox, 1997). What accounted for Houston's place at the forefront of the surviving urban growth frontier were new jobs, employment increases and, hence, population growth in the metropolitan area. The number of manufacturing and non-manufacturing jobs was significantly augmented between 1950 and 1982. As a consequence of the process of globalization, manufacturing in other large metropolises like London was in the process of dismantling, with the services sector moving in at an accelerated rate. In contrast, in Houston, manufacturing jobs increased about fivefold, from 59 200 to 254 000, while non-manufacturing jobs grew at an even faster rate, about sevenfold (from 180 000 in 1950 to 1 329 400) (Feagin, 1988). Houston was second in the US in both employment gains (80.7) and in rate of population gain. Houston's growth rate dropped in the 1980s, but by the 1990s the city had recouped nearly all the jobs it had lost in the recession.

Moreover, between 1950 and 1980, government quotas were set for imported oil, and this action raised oil and gas prices, with the effect of stimulating Houston's economy.[10] Houston had become the oil capital of the world. By the 1980s, the number of manufacturing plants in Houston had increased dramatically, and the industrial complex extended over 25 miles along the Houston Ship Channel. In addition, an intricate underground system, the so-called 'spaghetti bowl', consisting of several thousand miles of product pipeline, connects some 200 chemical plants, refineries, salt domes, and gasoline processing plants crossing along the Texas Gulf Coast and the city of Houston (Greater Houston Partnership, 1990b).

Paving the way to massive globalization, the relocation and reorganization of powerful oil firms has been a central response to the crisis of decreasing profitability registered in the previous years (Thrift, 1988, Chapter 1). Leading multinational oil corporations, such as Shell and Exxon, Gulf and Texaco, had located their major operating and other subsidiaries in this particular metropolitan area, and supported major research and development centres (Malecki, 1981). In the 1980s, Exxon built one of the world's largest industrial parks in Houston, with mostly major chemical firms as tenants, while Exxon's largest refinery is in Houston. In the global economy, Houston has since then appeared among the most specialized international cities in the energy business. Although the city has not housed most of the corporate executives who make the broadest worldwide capital investment decisions for those firms, Houston has a concentration of more organizational units of oil- and gas-related firms than any other city in the world (Feagin, 1988). It also has housed the headquarters operations and subsidiaries of smaller multinational oil firms. These offices often controlled oil operations across the US and in some cases across the globe.

The city expanded horizontally and vertically; with its seven business centres and hundreds of major plants, office towers and shopping centres, Houston had more than 1200 oil companies and supply houses: 'oil facilities, from refineries to office skyscrapers were the concrete embodiments of a continuing oil boom' (Feagin, 1985, p. 1214). The growth trends which have prevailed in Houston in relation to the other five largest US urban areas are illustrated by the continuous rise in capital invested in manufacturing, especially in refineries and petrochemical plants between 1967 and 1982 (see Figure 4.2). The present surroundings of large sections of affluent and also low-income residential areas in the east side of the city consist of industries and refineries. The extended industrial complex covers over some 25 miles (see Plate 1). In the 1990s, Houston had 3310 national and international plants with more than 2000 chemical plants alone (Greater Houston Partnership, 1990b). This represents an extraordinary increase from the some 420 manufacturing plants of all sizes found on the eastern side of the city in the early 1930s.

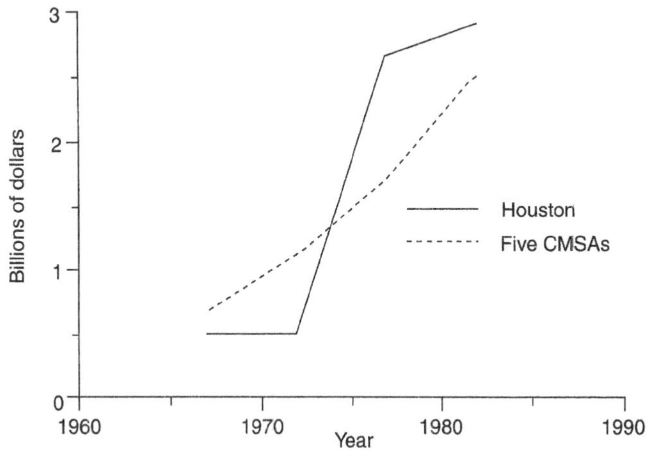

Figure 4.2 New capital investment in manufacturing: Houston and the mean for the five largest CMSAs,* 1967–82
*The five CMSAs are New York, Chicago, Los Angeles, Detroit and Philadelphia.
Source: US Census of Manufacturing, adapted from Thomas and Murray (1991), p. 50.

In summary, the natural and social processes that have affected the environmental quality of the city, and not least the health of the residents, as will be seen later, emerge as interconnected. This interaction is understood when approached by a view which considers the economy beyond the local level. This is so because 'changes in cities are a result of global and local capital as well as resulting from state policy at local and national levels' (Feagin and Smith, 1987, p. 17). The book has argued that while there are general tendencies or causal processes at work in capitalist societies, they do not operate in isolation, but in combination with historical, social and biophysical local circumstances (see Chapter 3). The result is 'variety and uniqueness with particular local combinations and development of the tendencies' (Massey, 1987, p. 120). The combination of global and regional processes and local events is key. We have integrated the historically specific analysis with our political-economy framework of current structures to respond to the context and change at global and local levels as specified in the multi-faceted model in Chapter 3. It has been established these are crucial

Plate 1 A close view of one section of the Ship Channel discloses the great density of establishments. More than 1200 hazardous installations operate along the 25-mile channel. Their dangerous presence and massive release of toxic emissions reflect frenzied and uncontrolled industrial growth and complete disregard for environmental and human consequences.

aspects to explain persistent current air pollution in world cities. Economic growth as well as the globalizing tendencies of capitalism have been the most influential processes in causing severe environmental degradation in Houston, and this has been supported by the generative, rather than reactive, role of the government in economic growth. Capital influx has been unrestrained, with the absorptive capacity of Houston environmental systems quickly overwhelmed (Cherni, 1993a). The next section moves on to examine the contradictions between growth and air pollution over time. This is needed to give the theoretical and contextual issues of growth and globalization an empirical perspective on events that is crucial to building knowledge.

The contradiction of Houston paradise and Houston hell

There has been especially widespread urban expansion and population growth in world cities. In the case of Houston, these have taken place since the beginning of the twentieth century (Feagin, 1988; Shelton et al., 1989; Thomas and Murray, 1991). However, since the 1950s in particular, residential, commercial and industrial developments have extended across both the city and suburban boundaries without any concern for planning land use. Such trends brought remarkable changes to the biophysical landscape (Thomas and Murray, 1991) and in the quality of the urban environment. These changes have created typical American landscapes, commonly seen in Houston and in other cities: landscapes made to suit vehicular transportation, suburban housing and shopping centres, 'a sameness in commercial strips everywhere' (Jackle, 1994); luxury condominiums which reflect the 'landscapes of private power and wealth' (Wyckoff, 1994, p. 235); and a landscape of 'specialized activity and mechanical integration, of growth and decline, and of abandoned and reused relics' (Meyer, 1994, p. 249).

Rapid and unplanned economic growth has taken place in Houston. Epithets such as 'the pearl of the Sunbelt', 'the capital of energy', 'the free enterprise city', 'the oil city', 'the miracle city', and 'the space city' (due to the space centre NASA) reflect the particularly dominant economic character of the city. A colossal petrochemical complex, including refineries and industry, is located in the Houston region.

A further point is that the medical sector of Houston stands out because of its size and quality. The Texas Medical Center in Houston is the largest medical complex in the world and the main employer in the city (Greater Houston Partnership, 1995/96). Health care is one of the region's most important 'industries' and the medical facilities in the city are numerous and impressive. This world city is also internationally renowned for its cancer treatment and research centre, The University of Texas M.D. Anderson Cancer Center, for top medical schools, such as the Baylor College of Medicine, and for pioneering new methods in surgery and heart disease treatment.[11]

Despite the idyllic image that this urban space, with its indisputable thriving economy and medical excellence, might project, it bears intense contradictions that position the achievements of growth in stark contrast to environmental and human failures. Indeed, awe, admiration and also disgust for the city cannot be avoided because Houston's current built environment is impressive, unique and grotesque at the same time. Amidst a flat and dull landscape stands a seemingly impenetrable conglomeration of high-rise buildings. This is the central business district, or downtown. The flat surroundings seem intimidated by the presence of this cohesive mass of skyscrapers, many of which exceed sixty floors in height. A few other isolated groups of very high buildings can be distinguished on the city's horizon (see Plate 2). Here and there 'a shopping center disrupts the pattern by its scale' (Jackle, 1994, p. 307). The dominant character of these high rises is accentuated by the fact that many of them are built wholly of blue, turquoise or brown glittering glass. The Houston Medical Center presents distinctive characteristics of affluence. The presence of numerous highways is another prominent feature. Very wealthy neighbourhoods of mansions and tree-lined boulevards are found in Houston; such social and spatial exclusivity is characteristic of many American cities (Wyckoff, 1994). Elkin et al. (1991) have pointed out that 'nothing enhances a city's reputation more than friendly streets where there is plenty happening' (p. 13).

Plate 2 The Transco Tower, situated in a popular commercial sector of the city (the Galleria), is the site of oil-related offices and the tallest building in Houston. Looking like a chocolate bar, as it is popularly known, it is simply an eyesore on the landscape. Impressive height and ostentatious presence of the Transco Tower epitomize the embodiment of capitalist growth, flourishing business and powerful multinationals in the city.

But not all that glitters is gold. The horizon of the city seems to expand forever with no interesting vestiges of nature to attract the observer's attention. Along main roads and highways, one can see derelict and rusting cars, abandoned buildings, empty land and poor houses. Conditions in some low-income neighbourhoods are rundown, with decaying flats, both old and newly built, minority quarters, and commonly such eyesores are in the vicinity of well-off residential areas. Weather conditions in Houston are also quite extreme. High temperatures and unbearable humidity reign for more than six months of the year; average temperatures range between 56 °F and 76 °F, with a relative humidity of 76 per cent (MacDonald, 1976). To facilitate life in this climate, most buildings are air-conditioned. Houston's population cocoons itself from the heat within artificially air-conditioned buildings; indeed air conditioning has played an important role in the city. Hot weather in Houston usually starts in May and lasts until October. Other big cities have higher summer mean temperatures, but few have the humidity levels of Houston. Office air conditioning first appeared in 1923, but most business locations were not cooled until after the Second World War. Central air conditioning was an essential factor in attracting business to Houston. Homes, cars and schools were air-conditioned in the 1950s and 1960s, followed by other specialized sites in the 1970s. By the mid-1970s, the middle and upper classes had almost completely insulated themselves from the four to five months of severe summer heat and humidity (Thomas and Murray, 1991). Within Houston's extended city, mobility is predominantly by private car, congestion is commonplace and public transport is very deficient.

Unpleasant smells from oil and gas refineries usually invade the city (see Plate 3). There are major problems with routine sewage and garbage disposal. Unique problems of subsidence affect the city. Many areas are actually sinking as a result of vigorous development, and flooding problems have increased through subsidence (Feagin, 1988). Water supply is inadequate, and there is persistent water and air pollution. In Houston, outdoor air conditions are usually very unpleasant. In addition to high temperatures and humidity, emissions from numerous cars and industry, and the heat produced by central cooling equipment, contribute to at times unbearable air quality. Houston's remarkable cancer mortality and high rates of

Plate 3 What could be a 'pure' pastoral scene in the San Jacinto Monument Park is abruptly disrupted by the presence of a solid background body of chemical plant. Oil- and gas-processing facilities are located within easy reach of main roads and the landscape these create has become typical of the east side of the city.

infant, childhood and maternal mortality (Chapter 6) are all the more remarkable if we consider that the city is ranked among the ten wealthiest areas in the US in terms of personal income (*Houston Chronicle*, 1990b).

One of the most obvious contradictions is that high levels of air pollution make Houston one of the most contaminated cities in the US (*Houston Post*, 1990a; TACB, 1992c; US EPA, 1991). To some extent Houston resembles other American cities in experiencing such high levels of general contamination like Los Angeles, Detroit, St Louis and Chicago, yet the emissions from industry in particular have reached alarming levels in the city, as shown in the next chapter. A key issue for this urban pollution is that industrial emission tends to be localized and, despite the fact that large sections of Houston have been frequently invaded by industrial pollution, the residential areas located nearby petrochemical plants have been most adversely affected (see Plate 4). Because the industrial

Plate 4 A view that shows integration of expansive industrial development and housing in a low-income street in the Ship Channel area, indicating an evident source of urban contamination from nearby chemical plants.

development is firmly linked to the international economy, the contamination found in large areas of the city represents a local manifestation of a global condition. Another contradiction is that access to competent specialized and abundant medical facilities offered in Houston, as anywhere in the US, is very expensive and usually quite beyond the means of those without adequate finances. Not surprisingly, the standard of health among disadvantaged residents is poor, and indicators of public health have rated deficient in relative terms. Despite the fact that Houston has ranked among the most air-polluted cities in the country, with anxiety about water quality being endemic, any condemnation of these living conditions as endangering residents' health expressed by the city's medical and political institutions has been studiously ignored.

Powerful economic institutions, abundant wealth and resource-rich medical services in Houston have stood in stark contrast to the deplorable state of the environment in many sections of the city and the limited institutional protection for citizens in the face of imminent risk. The paradox reflects the elemental contradictions of growth. Through industrialization, urbanization and globalization, most major cities of the developed world experienced influential changes in the twentieth century. In this light, Houston offers itself as an appropriate case for highlighting some of the main afflictions within economically developed cities at the end of the twentieth and beginning of the twenty-first century.

Houston has been given many positive epithets as a consequence of economic success in the global economy, the 'oil capital of the world' perhaps being the most widespread and powerful representative. It is now, however, time to make connections between this period of economic expansion in the energy sector and its concomitant privileged position in the international economy, and the local environmental degradation that took place during the same period. Such intensive economic activity resulted in severe ecological problems which were already evident in the early twentieth century; they continue unabated to this day.[12] The health of the population is not something of which this world city can be proud.

The discovery in the early twentieth century of oil in Houston and its surroundings brought significant pollution, greatly increasing the region's environmental problems. Clearly, 'in the rush for instant wealth, oil was seen as black gold, not black sludge' (Pratt, J.A., 1980, p. 227).[13] By the late 1940s, cotton accounted for only 10 per cent of exports through the Port of Houston, while oil accounted for 80 per cent. Oil and gas industries had surpassed agriculture as Houston's dominant economic sector.

Speed of extraction and quantity of production – rather than efficiency and the prevention of waste – characterized the early period. Although the exact levels of air and water pollution in the Houston area could only be roughly estimated, it is clear that environmental pollution was a problem at a very early date. Several days after the first strike in 1901, the Spindletop well, massive pollution followed: a thick, yellow, sulphur-laden fog began periodically to discolour houses and threaten the lives of some of the drillers (Pratt, J.A., 1980). Oil extraction often led to massive drain-off of crude that soaked the ground; the rapid removal of oil from the wells also resulted in the introduction of salt water into the underground reservoirs of oil and into the region's water system. For lack of anything better, producers often used unsafe open earthen pits and wooden tanks for storage. Adequate transportation and loading facilities did not exist, so large quantities of oil were lost between the pumping stations and the tankers that carried much of the crude to the East Coast. Earlier, coal-burning trains had brought unhealthy smoke to Houston; the rapid clearing of timber had stripped the soil of its protective covering; and the milling of this lumber in coastal towns like Beaumont had resulted in contamination of Houston's air. The growing refining complex on the Gulf Coast caused considerable pollution, but for most of the century no institutions – neither public nor private – had sufficient power or incentive to deal effectively with the resulting problems (Pratt, J.A., 1980, p. 225). Pratt adds: 'It is likely that at least as much oil found its way into the region's ground, water, and air in this period as found its way to market' (p. 228).

Oil production on the Gulf Coast, and in Houston in particular, increased dramatically between 1929 and 1941. Most prominent was the great East Texas field, discovered in 1931, which in less than ten years multiplied about eight times the number of producing wells in the Houston area, from 3612 in 1932, to 25 765 in 1939 (Williamson et al., 1963). It is still among the most productive in the US (Thomas and Murray, 1991). Oil production in the Gulf Coast area continued to increase fourfold annually between 1929 and 1941, from 57 to 226 million barrels (Williamson et al., 1963).

Before the government started regularly to monitor air pollution in the 1970s, the first environmental survey in the region in 1923, the *Pollution by Oil of the Coast Waters of the United States*[14] reported that the Houston Ship Channel, where the petrochemical works and the port are placed, represented 'one of the worst oil polluted localities seen by the Committee' (Pratt, J.A., 1980, p. 234). More than 20 years later, in the 1956–58 survey of the Houston area, it was officially disclosed that the 'air pollution problem resembled that of Los Angeles', which had captured the nation's attention since the late 1940s (*Air Pollution Survey of the Houston Area, 1956–1958*, cited in Pratt, J.A., 1980, p. 243). Industry was cited as the biggest polluter in Houston, with municipal sewage a major secondary source. The follow-up survey in 1964–66 revealed how quickly the limits to natural pollution dispersal could be reached in Houston in a period of rapid growth. The conclusion to be drawn is that the findings of these later environmental reports were no less distressing than those of the first national pollution survey of 1923, which had acknowledged for the first time the severity of contamination caused by economic growth.

Major environmental problems have existed in the Houston area as a result of failure to deal adequately (if at all) with industrial waste from petrochemical and primary metals production. In addition, municipal waste produced by a growing population, and car emissions that accompanied the expanding use of the automobile, made Houston one of the most seriously polluted areas in the US. The economic activities that gave rise to such environmental destruction have not only continued to operate but have also multiplied, so that concomitantly, further environmental degradation followed, although production techniques and product lines have changed dramatically over the years.

In 1990, the results of the first *Texas Environmental Survey* (Klineberg, 1990a), which collected people's opinions on regional environmental problems, once more highlighted the pervasive presence of air pollution in the Houston area. Furthermore, in 1994, the EPA national records of air pollution had positioned Houston as the second most air-polluted city in the US. By the year 2000, the affluent city stands just by the side of Los Angeles as the real capital of pollution in the United States.

Conclusion

The tendency has existed to depoliticize environmental issues at the international level, while considering resource conflict at the local or national level as other than environmental. Therefore, crucial to any analysis of local environmental degradation, such as urban air pollution, is the identification of the participating political and economic structures. Social structures deploy causal powers and actively encourage – or not – the creation of wealth. These do not exist in a vacuum but connect to recent historical developments. Against this background, the internationalization of the local economy and the environment took place in Houston. Hence, economic growth has been at the root of major environmental changes.

The relationship between society and environment in Houston has been very contradictory. While economic growth has created the conditions for the proliferation of massive wealth and prosperity, it has also seriously damaged the natural environment. Hence the population has remained at risk from ill health due to the consistent presence in the city of contamination originating in the industrial operations that achieve growth. The second contradiction of capitalism underpins the contrast found between wealth and risk.

Following some of its theorists (for example, O'Connor, J., 1994, Ch. 1), the second contradiction establishes that nature is being capitalized, that is, it is given a commercial value, or it is commodified as part of the drive to produce and reproduce capital. In Houston, the capitalization of nature became not only a frequent feature but also a grotesque characteristic of the way that society interacts with nature. In the oil business, the useful configuration of oil, gas, sulphur and water has been hideously exploited for profit-making purposes. The local air, water sources and soil have been used as depositories for the enormous quantities of toxic by-products emanating from industrial processes.

This chapter has identified the particular aspects of the global and regional economy involved in the rise of local pollution. It has singled out the ways that degradation has taken place over time. The chapter has shown that the 'something' adversely affected by society, the issue raised in the Introduction, has not been difficult to identify in the world city of Houston: the quality of its air, water and soil has been the most adversely affected aspect of nature. The contradiction between economic growth and environmental contamination has emerged simply because basic natural resources needed to operate industrial processes have been severely abused and largely unprotected. Environmental considerations would have meant a diversion from the fundamental growth objectives that spurred the 'treadmill' which has enabled the reproduction of further growth (Schnaiberg and Gould, 1994, Ch. 3). There is, however, an additional point to be made here. Paradoxically, the contradiction between economic growth and the regional environment is possible because of an additional factor. The contradiction also serves to indicate that there exists a tight interconnectedness, rather than opposition, between the social and the biophysical dimensions that coexist in cities. This recognition overcomes the particular cognitive dualism between nature and society. It is important to acknowledge concrete aspects of the working of the second contradiction.

Certainly, Houston was a highly economically developed city before the expansion of the energy sector, and environmental pollution was also known before the escalation of the globalization era. However, the ecological disruption that the nationally supported petrochemical industry and internationalization of the economy brought to the region since the 1960s and 1970s has been radically different and many times more severe than that of the first half of the century. Undeniably, the character of the world corporate city of Houston in the 1990s was potentially dangerous, and particularly so for residents of the East Side.

This chapter has shown that while familiar trends of increasing economic growth and globalization have affected many regions in the world, the implications for both a city's environment and its population are substantially different, depending to a great extent on local historical, political, economic, institutional and climatic factors. In Houston in particular, the US centre for its hitherto largely unregulated oil industry, the strategies of growth have, over time, shown scant regard for the human and environmental element in the economic equation, initially from ignorance and lately for economic reasons.

Throughout the twentieth century, Houston became increasingly important as a world energy-related centre with consistent population growth and urbanization mirroring changes in the economy beyond the city's own boundaries. Tens of miles of chemical plants, impressively tall glass buildings, luxurious facilities for a population accustomed to shop and consume, are among some of the most visible outcomes of economic growth. A less visible but no less significant output has been air pollution. Following the rationale of the 'transformative approach' (Chapter 3), in Houston the finite absorptive capacity of the biophysical environment and the natural threshold of humans to withstand the impact of industrialization have long been ignored in the rush for profit-centred activity. The ultimate consequences of this type of growth have been substantial employment opportunities, accelerated and uncontrolled urbanization, undeniable massive private wealth for a few, and hazardous contamination for all.

It is apparent that globalization and regional economy intimately intertwine in the local space of world cities. The process of globalization and local economic growth is grounded on a subsidiary relationship between the two, whereby each one needs the other to continue advancing this type of economy. This relationship has also been functional, as it has fostered capital accumulation on a large scale, particularly in the energy sector, so enhancing the oil economy. Local economic agents who recruited financial and political support to maintain lax environmental regulation in order to promote a competitive business climate in the city have favoured economic growth in Houston.

The next chapter moves on to address the current state of the quality of the local air. The state of the air seems to reflect a straight continuation of the previous conditions as discussed above. Chapter 5 will begin to enquire into the levels of ill health found in the city of Houston and into how these connect to structural and atmospheric conditions.

Notes

1. Given the volume and weight of petroleum, Child Hill and Feagin (1987) argue that there were obvious advantages to refining it near production sites. The refining process requires large quantities of fresh water and access to port facilities (or, later, to pipelines) that can move the finished product to market.
2. One needs to consider, however, that the power of large corporations and transnational banks is insufficient to explain their capacity for global control. The practice of global control, the work of producing and reproducing the organization and management of a global production system and a global marketplace for finance constitute the capability for global control (Sassen, 1991).
3. The decision to expand a refinery was taken in New York or Pittsburgh. The determination of regional wage rates was influenced by negotiations between union headquarters in Denver and national and international officials of the oil companies. From Washington came decisions affecting a wide range of refinery-related matters, from the size of the owner company to the volume and type of effluent discharges allowed (Pratt, J.A., 1980).
4. For example, in the 1940s, the national government embarked on a crash programme to develop a huge synthetic rubber industry: 'not surprisingly, the Houston Ship Channel was selected as the primary location for this new petrochemical industry' (Thomas and Murray, 1991, p. 46).
5. Shell Oil located its US administrative headquarters there; Exxon concentrated more administrative and research operations; Gulf, Texaco, and Conoco located or expanded major national subsidiaries in Houston (Feagin, 1985).
6. Houston had long been the world centre for oil technology. In the 1920s this meant little more than developing better drilling tools, but by the 1950s the industry was concerned with massive tasks such as designing and building off-shore drilling platforms (Thomas and Murray, 1991, p. 47).

7. These funds allowed the channel to be widened and deepened, so that after 1914 ocean-going ships could use the port without reducing their cargoes (Thomas and Murray, 1991).
8. The Houston Metropolitan Consolidated Area (CMSA, 7422.38 square miles) consists of three PMSAs (Primary Metropolitan Statistical Area), Houston, Galveston–Texas City and Brazoria.
   The Houston–Galveston–Brazoria Consolidated Metropolitan Statistical Area's (CMSA) population (3.7 m) ranked tenth among the country's metropolitan areas, and was the largest in the South and Southwest of the US.
9. The impact of the world restructuring in the oil economy and in the manufacturing sector in Houston was profound (Feagin, 1988). The 1982–87 crisis and reorganization in the world oil–gas industry were linked to 10 per cent unemployment, over 1000 bankruptcies, oil refinery and other oil company lay-offs, and cutbacks in the petrochemical industry (Child Hill and Feagin, 1987, p. 174).
10. Another important example of federal intervention was the decision to build the National Aeronautics and Space Administration complex (NASA) in Houston. Land for the NASA complex had been donated by Humble Oil, Exxon (Feagin, 1985).
11. The Texas Heart Institute was the site of the first human implant of the air-driven left ventricular assist device and one of the first artificial heart transplantations was carried out in Houston (by M.E. DeBakey).
12. The 'lack of reliable, systematic sources of information on the extent of pollution in the early period' (Pratt, J.A., 1980, p. 249) makes it necessary to rely only on two main historical sources, J.A. Pratt (1980) and Williamson et al. (1963).
13. Pollution problems from oil refining fall into two main categories. First, pollution arising from the production process – drilling, transport, refining (including accidents and explosions) – and the petrochemical industry refineries emitting toxic hydrocarbon vapours, combustion waste gases, sulphur-containing gases, and fine particles; and second, problems arising from the consumption of oil, e.g., car emissions, and the so-called 'greenhouse effect' or global warming, and from accidents (Strauss and Mainwaring, 1984).
14. This was a comprehensive investigation of national water conditions. The American Steamship Owners' Association and the US Bureau of Mines cooperated in completing the survey (Pratt, J.A. 1980).

5
Monitored and Reported Local Air Pollution

Introduction

A concrete legacy of rampant industrial development over time and the integration of the city's economy particularly the energy sector, in the international market are evident in a world city like Houston. A disfigured landscape of oil refineries, petrochemical plants and other industries extends over some 25 miles long. Remarkable levels of pollution could not but arise from years of activity in such a massive and concentrated industrial complex. This chapter shows that air pollution has remained as uninterrupted and striking a feature of Houston as has economic growth. The previous chapter traced the origins of environmental pollution to the development of the oil industry in particular. Local industrialists, national policy and the global economy have been identified here as the main promoters of the oil-led growth in this part of the world. In the past, official emission control has suffered from a lack of commitment towards the environment and the residents. Identification of the chain of interactions between social developments and environmental degradation in one city contributes to challenge the well-established dualism between the interests of the natural and social sciences. It is argued now that, while the previous chapter showed that this history of growth and success has gone hand in hand with abusive extraction from nature or environmental withdrawals, this chapter sets out to prove that the other corollary has been the toxic additions to the environment. Analysts have usually failed to connect these pieces, and have separated what is fundamentally inseparable, namely, unregulated growth and the costs in terms of damage to nature and people. This chapter analyses the extent of local and regional air pollution. It combines this analysis with assessment of attainment levels of US health safety standards limits in recent times. Public health data on cancer death and residents' report of child ill health complement the technical data.

The natural sciences have revealed the mechanisms whereby toxic emissions can damage the environment, and the health impact of deteriorated environments on the population has been increasingly acknowledged. As previously stated in Chapter 1, it is a complicated matter to define air pollution. Large amounts, and a yet unknown variety, of toxic chemicals are released every year into the atmosphere by industrial processes, power stations, traffic and other users of fossil fuel. Combinations of such materials with atmospheric conditions and their levels of concentration vary. Although chemical reaction is the crucial event in the formation of urban air pollution, additional factors also affect the degree of pollution found in cities. Three basic conditions are required for these chemical emissions and chemical formations to degrade the environment and to become health hazards: emissions must exceed the natural capacity of the environment safely to absorb them; next, they must engage with climatic and topographic features (for example, sunny conditions may precipitate a photochemical reaction if the concentration of pollutants is high; atmospheric inversion may contribute to a pollution event; pollutants released in valleys may remain longer in the air); finally, they must combine with other chemicals to produce new contaminating materials. Although 'proximity' is the most widely used parameter for assessing exposure, pollutants also travel, mix and create new pollutants, and therefore certain types of pollution may be found in areas which are not industrial and are far from the source of emission. This chapter gives a comprehensive account of recent air pollution in Houston.

The information on the combination of atmospheric, ecological and chemical mechanisms is crucial, but still it is not sufficient to explain widespread urban air pollution and subsequent ill health. Moreover, such conditions may not be able to give rise to urban pollution. Technology and political institutions play central roles in determining the environmental conditions prevailing in a particular society. That is, the presence – or not – of air pollution depends in great measure on, for example, the type and quantity of legally permissible emissions, the fuel combustion technologies in use, the allowed motor travel speed, rules and devices that control industrial emissions, and the applicable pollutants' health safety standards. Climatic conditions may only be found in specific geographical areas, affecting a whole range of local factors and residents, hence the importance of considering spatial distribution of pollution in order to assess which factors may most accurately be associated with ill health.

Put simply, the three preconditions do not alone produce the air pollution, which can trigger considerable ill health. When certain toxic chemicals from industry are released, certain climatic conditions such as inversion of cold and warm air may exacerbate the reaction of the toxic emissions and so produce considerable environmental pollution. Equally, the presence of pollution in the environment will produce ill health in people who live under and/or are exposed to certain conditions. For example, air pollution may stimulate illness in children who have been exposed to pollutants. The risk of such exposure may well increase under suboptimal spatial and socioeconomic circumstances. Therefore population exposure to air pollution, while applicable to all, is not applicable in an unqualified or unspecified way. Hence the relevance of both monitoring pollution and surveying residents in world cities.

Local residents provided information crucial to determining how social and biophysical factors interact. This was necessary to establish whether, and in which socioeconomic and geographical circumstances, ecological mechanisms are activated to cause ill health. This chapter answers the question of the physical circumstances. The database consisting of residents reports of their accounts and experience of local air pollution and ill health was analysed and cross-referenced. The chapter suggests a correspondence between the information collected by the government agencies' monitoring systems and residents' reports, and this issue will be taken up in the following chapters. The cluster design of the questionnaire and the comparative nature of the case study's large representative sample provided essential quantitative and qualitative information from which to challenge some assumptions.

The failure to meet environmental targets

In the last three decades four different agencies simultaneously monitored and controlled air pollution in Houston – the federal US Environmental Protection Agency (US EPA), the state Texas Air Control Board (TACB), the City of Houston Bureau of Air Quality Control (BAQC), and the private Houston Regional Monitoring (HRM). However, they have apparently not ensured noticeable improvements in the quality of the city's air. Ozone, the main current pollutant found in Houston, has become a problem of the last decades in most world cities. For example, in London, ozone concentrations have increased between two- and threefold (ENDS, 1994b). Ozone is a necessary and desirable component when it is in the upper atmosphere at heights of five to ten miles above the earth's surface, where it partially blocks the sun's radiation. However, at ground level, high ozone concentrations in urban areas constitute a contaminant and is the most difficult and expensive air pollutant to control (TACB, 1987). A number of factors contribute to this difficulty. Unlike other gaseous pollutants, ozone is not emitted directly into the atmosphere. Instead, it is created by the action of sunlight on volatile organic compounds and nitrogen oxides, mostly deriving from fossil-fuel combustion. In Houston, ozone excess is the main air pollution problem and it has been the main cause of failure to attain the six National Ambient Air Quality Standards (NAAQS; see Table 5.1). Under the national specifications, an air pollutant episode is described as any measurement that is higher than any of the six NAAQS irrespective of any other specification. 'Attainment' and 'non-attainment' levels are the criteria that reflect the degree of compliance with the nationally established NAAQS. Assessment of air quality under the national rules tracks two kinds of trend, air concentrations and emissions, and incorporates a broader range of meteorological conditions and control strategy considerations (US EPA, 1991).

Primary standards are designed to protect public health; secondary standards to protect public welfare, that is, vegetation, materials and visibility.[1] There are many more pollutants which are recognized as having carcinogenic effects but for which national limits have not been set. Such are the so-called volatile organic compounds (VOCs), which are composed of 157 toxic pollutants and are

Table 5.1 National Ambient Air Quality Standards (NAAQS)

Pollutant	*Average period*	*Primary NAAQS*	*Secondary NAAQs*
PM-10	Annual*** 24-hour*	50 ug/m^3 150 ug/m^3	Same as primary Same as primary
PM-25	24-hour	65 ug/m^3	Same as primary
Sulphur dioxide	Annual*** 24-hour** 3-hour**	0.03 ppm 80 ug/m^3 0.14 ppm 365 ug/m^3	 0.50 ppm
Carbon monoxide	8-hour** 1-hour**	9 ppm 35 ppm	No secondary standard No secondary standard
Nitrogen dioxide	Annual***	0.053 ppm	Same as primary
Ozone	Maximum daily 1-hour* average	0.12 ppm	Same as primary
Lead	Maximum quarterly*** average	1.5 ug/m^3	Same as primary

* Not to be exceeded on more than three days over three years.
** Not to be exceeded more than once per calendar year.
*** Not to be exceeded.

Source: Adapted from US EPA (1990b, p. 2-2) and from Texas Air Control Board (1993a, p. 2).

emitted from sources as diverse as automobiles, refineries, chemical manufacturing, dry cleaners, paint shops and other sources using solvents (TACB, 1992b). However, the TACB uses estimated guidelines for many non-regulated chemicals.

In the US, the ozone problem may have had a long association with southern California, but it is a very serious problem in other regions such as the Texas Gulf coast. Had it not been for the US EPA relaxation of the ozone standard from 0.08 to 0.12 ppm in February 1979, 'following strong pressure from the petroleum industry, many more tens of millions of people would be regarded as living in areas with unacceptable air quality' (Elsom, 1992, p. 221). It should be stressed that the US EPA has proposed lowering the acceptable ozone standard to 0.08 ppm and extending monitoring time from one to eight hours (Arsen and Darnay, 1994). Houston has ranked first among seven Texas MSAs which violated national standards for ozone (see Figures 5.1 and 5.2); moreover, the city ranked second in the US among 60 of the US MSAs to exceed the ozone NAAQS.

Levels of ozone rise during sunny days with light wind speeds, and ozone trends are strongly influenced by annual variations in meteorological conditions.[2] Moreover, ozone tends to be an area-wide problem, with fairly similar levels of concentration occurring across broad regions. When focusing on its health effects, one should bear in mind that ozone is not simply a localized hot-spot problem. For example, Chapter 2 explained how children were at

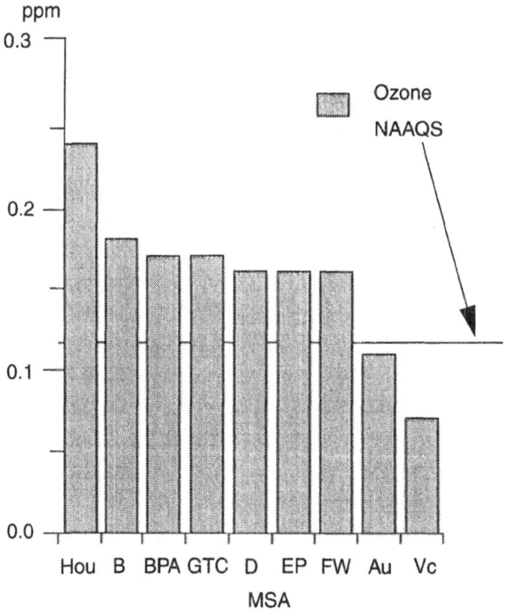

Figure 5.1 Comparison of peak ozone measurement* for Texas metropolitan and regional areas,** 1990

* Each bar shows the highest measurement recorded at any one site in each of the areas shown.

** Hou: Houston; B: Brazoria; BPA: Beaumont–Port Arthur; GTC: Galveston–Texas City; D: Dallas; EP: El Paso; FW: Fort Worth; Au: Austin; Vc: Victoria.

Source: Data from TACB (1993a), p. 4.

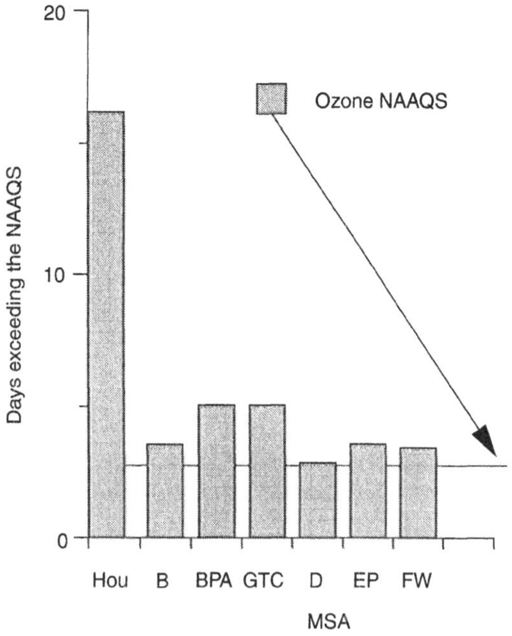

Figure 5.2 Comparison of highest number of ozone exceedance days for seven Texas metropolitan and regional areas,* 1988–90
* Hou: Houston; B: Brazoria; BPA: Beaumont–Port Arthur; GTC: Galveston–Texas City; D: Dallas; EP: El Paso; FW: Fort Worth.
Source: Data from TACB (1993a), p. 4.

higher risk of developing respiratory problems in Mexico City when they were exposed to ozone pollution peak, > 0.13 ppm, for two consecutive days in 1988 (Anderson et al. 1997; Romieu et al., 1993). We also mentioned how chronic cough, bronchitis and chest illness in both children and adults were registered in conjunction with a high count of particulate pollution (Schwartz et al., 1991). There is accumulated medical evidence to show that high levels of atmospheric ozone exacerbate pre-existing respiratory disease, reduce lung function, and cause increases in emergency room attendance, admission to hospital and mortality (see Chapter 2).

Houston was the second most polluted US city in the 1990s after Los Angeles to exceed permissible ozone concentration levels. It had registered a maximum hour concentration of 0.22 ppm when 0.12 ppm is the permitted NAAQS (BAQC, 1991), albeit the city was said to be a very distant second with only 52 days in excess compared to 137 days in Los Angeles (McMullen, 1989; TACB, 1992a). None the less, when compared to cities with the next highest levels of ozone, it is evident that the latter are very far below Houston: Philadelphia with 17 days; Detroit with 10 ozone excess days; and New York with a maximum of 5 days (data from US EPA, 1991, table 5-3, p. 5-4). Recognition that the ozone problem has definitively been worse in California than in any other US state does not invalidate the fact that even when pollutant measurements are smaller, affected areas can be severely distressed.

A valuable source of information for putting ozone contamination in spatial and historical perspective is the US EPA record of regional air pollution. Air pollution reports for the Houston area reveal strikingly high levels of ozone, particularly near the refineries in the Ship Channel (see Figure 5.3). Also, in the Southwest area of

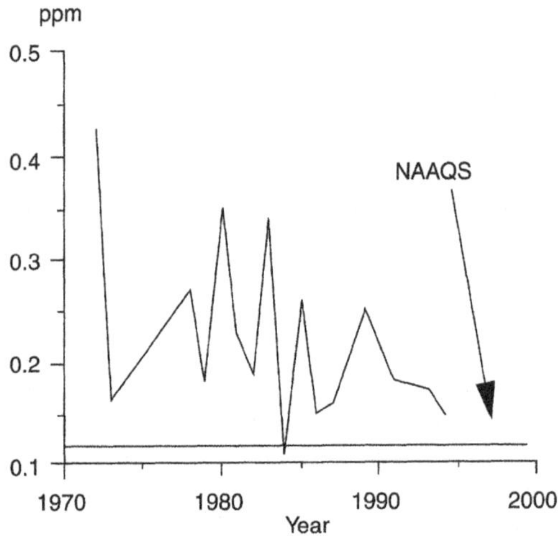

Figure 5. 3 Measured ozone levels* in the Ship Channel area, Houston,** 1972–94

*Ozone NAAQS = 0.12 ppm; **Clinton Drive site (see Figure 5.1).

Source: Data from US EPA AIRS.

the city, ozone levels exceeded significantly the national health guidelines, but not by as much as in the Ship Channel area (see Figure 5.4). Industry and motor vehicles have been identified as the main sources of atmospheric pollution in Houston, most of which is made by the action of VOCs, oxides of nitrogen and sunlight. In fact, refineries, petrochemical and other industrial plants are responsible for 54 per cent of all VOC emissions in Houston. Mobile sources contribute 31 per cent, while small localized sources (such as gasoline retailers and dry cleaners) emit 15 per cent of total emissions (McMullen, personal communication, February 1992). This association is of fundamental importance, for it points to the clear influence of industrial sources on the high concentration of pollutants in Houston. Furthermore, the industrial origin of persistent ozone pollution in highly transited areas in the city has been shown in an unpublished work carried out by the City of Houston BAQC (1991). Over seven days per week, in a ten-year period between 1980

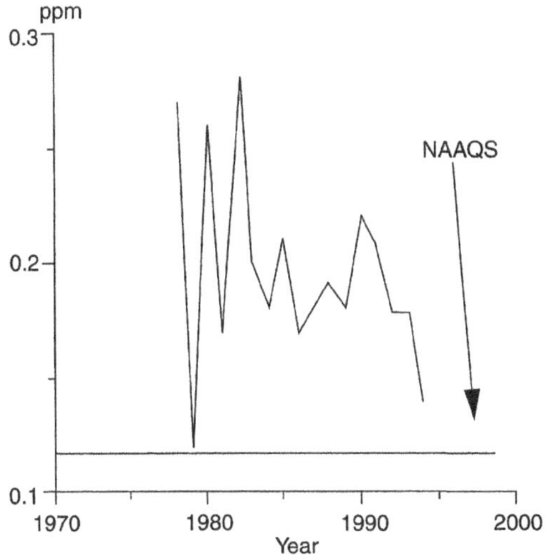

Figure 5.4 Measured ozone levels* in the Southwest area, Houston,** 1978–94

*Ozone NAAQS = 0.12 ppm; **Croquet Drive site (see Figure 3.2).

Source: Data from US EPA AIRS.

and 1990, the study compared the average number of days which exceeded the permitted ozone levels and the amount of traffic on a busy freeway located far from the industrial area in the Ship Channel. The study found that, while traffic count decreased significantly during Saturday and Sunday, the ozone average excess during those days remained as high as during the rest of the week. The only reason for air contamination to remain as severe at weekends as during weekdays was that oil refineries and other manufacturing plants in the Ship Channel operate on Saturday and Sunday. Industrial emissions travelling from the Ship Channel area were the source of persistent high ozone levels during weekends. The same source obviously contributes to the high concentration of ozone during weekdays (McMullen, personal communication, 1992).

In addition to ozone pollution, historically Houston has shown the highest measured sulphur dioxide levels in the state of Texas, despite the fact that no sulphur dioxide episodes have been registered there (TACB, 1993a). None the less, high sulphur dioxide levels in two sites in the Houston Ship Channel area had been recorded by private network HRM (TACB, 1992c, p. 5). As a result of high levels of sulphur pollution, a portion of the Houston Ship Channel area was considered for designation as one of non-attainment for sulphur dioxide as well (TACB, 1992b). Sulphur dioxide is produced by the burning of sulphur-containing fuels, from the smelting of metallic ores containing sulphur, and in the process of removing sulphur from fuels. In Houston, these processes provide for the global, national and local markets and take place in petrochemical plants located across the Ship Channel. Examination of historical annual records by monitoring sites revealed that the levels of sulphur dioxide had been much higher in the east, that is, in the industrial and refinery area, than in the southwest side of the city.

As to oxides of nitrogen, several types are produced by high-temperature fuel combustion, but only nitrogen dioxide has national standards. As a result of this limitation, the levels of nitrogen dioxide measured in Houston cannot reflect the real level of oxides in the atmosphere. Thus, levels of nitrogen dioxide have consistently registered well below the NAAQS levels (TACB, 1992b, p. 3).[3]

Analysis of the 1980s and 1990s reports strongly correlates with the results of the earlier surveys in 1923, 1956–58, and the 1964–66 survey, as seen previously. It indicates that overreaching the carrying capacity of the environment to absorb contaminants safely has continued to the point where there is significant deterioration in the quality of the air of this particular city. Furthermore, residents in Houston had reported the noticeable and unpleasant air pollution under which they had to live, as we shall see later in this chapter.

Uniform indicator of health risks

Relying only on ozone measures to assess environmental problems presents a number of difficulties. First, even if the exact effects of exposure to ozone could be scientifically identified, which they cannot, problems arise in determining the actual levels to which the general population may be safely exposed. Second, estimates can be made of the number of people who could potentially be exposed to the levels of ozone monitored each hour throughout the day. However, even on a day in the Houston area when numerous monitoring stations record high ozone values, there is a significant variation in the levels of ozone to which people might be exposed. Studies that consider spatial variability of concentration of pollutants are essential for the purpose of assessing exposure and understanding the severity of the health consequence (Chapter 2). Third, present monitoring machinery can measure pollutants up to a certain altitude only. As a consequence, concentration of pollution above the technical limit cannot be assessed (McMullen, personal communication, 1992). In this sense, the actual levels of concentration of pollutants and hence of exposure remain a mystery. Therefore, full reliance on ozone excess to assess the problem of contamination and hence, of health, may not be sufficient. Neither measuring nor modelling includes the necessary variables either to assess exposure or to assess health risks. To address this shortcoming, we shall now look at the Pollutant Standard Index.

The Pollutant Standard Index, PSI, simplifies the presentation of air quality data by producing a single dimensionless number ranging from 0 to 500, where 0 is good and more than 100 is unhealthy (see Table 5.2). The index is primarily used to report the daily air quality of large urban areas as a single number or descriptive word and it places maximum emphasis on acute human health effects. However, increased PSI levels generally reflect increased damage to the general environment (US EPA, 1993). The PSI uses

Table 5.2 The US Pollutant Standard Index and the quality of the air

Index range	*Description*
0 to 50	Good
51 to 100	Moderate
101 to 199	Unhealthy
200 to 299	Very unhealthy
300 and above	Hazardous

Source: US EPA (1991, p. 5-1).

data from all selected sites in the MSA and combines different air pollutants with different averaging time, different units of concentration and, more importantly, with different national standards. The PSI is computed for PM-10, carbon monoxide, ozone, nitrogen dioxide and sulphur dioxide (US EPA,; 1990b, 1991, p. 5–2).

Examination of US EPA records between 1984 and 1993 reveals a remarkably high number of days when the PSI in Houston was greater than 100. Compared to other major US metropolitan areas, Houston had the second highest PSI > 100. Figure 5.5 displays average number of PSI > 100 between 1984 and 1993 for

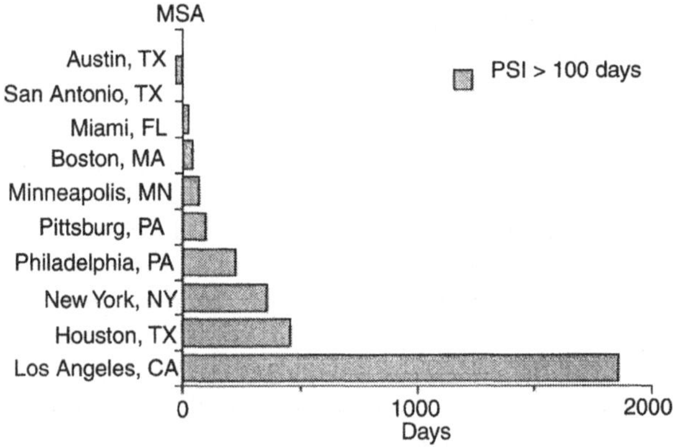

Figure 5.5 Cumulative PSI days greater than 100 in ten selected US MSAs, 1984–93
Source: data extracted from US EPA (1994).

ten selected MSAs. These cities were picked here to represent high, intermediate and low PSI > 100 days. There were 423 days with PSI > 100 over the period 1984–93 in Houston; at least 12 per cent of those were 'Very unhealthy' days. The highest number of PSI > 100 days occurred in 1987 and was 55. In Houston, 97 per cent of the PSI days > 100 were due to ozone excess.

Although most days in the year are clearly within the healthy limits, the cumulative number of unhealthy days in Houston in a ten-year period represents a source of concern. The city occupied the second place in the national PSI ratings, ranking below Los Angeles, CA, which, with 1851 days with PSI > 100 during 1984–93, was rated the most unhealthy city in the US. Houston with 423 days ranked considerably below Los Angeles; New York, with 359 days, and Philadelphia, with 224, also scored considerably less. These figures are illuminating when compared with cities with low number of PSI > 100, such as Miami, FL, 22 days; San Antonio, TX, 5 days; and Austin, TX, 2 days.

The PSI provides useful information to the public. It also serves the purpose of assessing cities by their overall environmental quality. Several problematic assumptions are implicit in the PSI analysis, however, the most important of which is probably that the monitored data available for a given area provide a reasonable estimate of maximum short-term concentration which may not represent the air pollution exposure for the entire area. If the downwind maximum concentration site for ozone is outside the MSA, this data area is not used in the PSI analysis, which thus fails to reflect geographical variations of air pollutants within one MSA. Local differences in pollution may be of utmost importance when studying its health impact. However, official assessment of people's actual exposure and subsequent health risk depends upon these measures. Finally, the PSI does not take into account the possible adverse effects of synergism, the effects of combinations of pollutants (US EPA, 1993), because each pollutant is examined independently. Combining pollutant concentrations is not possible at this time 'because the synergistic effects are not known' (US EPA, 1991, p. 5–7). This is likely to be a very important issue for health effects. The information provided by the PSI and the attainment criteria are essential to determine present levels of pollution in Houston.

Therefore, to overcome these further and significant shortcomings, we move on to discuss the information collected from local residents. This can indicate whether the pollutant standards themselves are adequate to safeguard the health of the population, and whether monitored concentrations and indexed information can reflect the risk effect on the population of particular geographical distributions of air pollution within one metropolitan area. Recognition of the apparent impact of pollution on health could throw light on the appropriateness of permitted levels of air pollution. This type of knowledge provides clues about the benefits and failures of relentless economic growth and globalization irrespective of environmental consequences.

Participatory information on air pollution and health

Some important findings derived from three different surveys – this book's *Air Pollution* and Child Health Survey (APCHS), 1990, the yearly *Houston Area Survey* (HAS), 1982–99, and the single *Texas Environmental Survey* (TES), 1990 – have indicated the following significant responses. They suggested that environmental quality in the city was very poor; that economic activity was directly implicated in causing prolonged air and water contamination; that the laws to protect the environment and to condemn polluters were insufficient; and that emission of recognized carcinogenic elements was highly feared. In different ways, the surveys neatly pointed to structural origins of environmental problems and highlighted the overall impact of pollution on the population. It was not expected that Texans would be overwhelmingly committed to environmental protection given the public pro-growth policy of the state, and the business atmosphere that has characterized this world city. However, the plain fact that, according to the HAS, 30 per cent of the interviewees in 1999, 36 per cent in 1995, but only 27 per cent in 1988 regarded the city's efforts to control air and water pollution as inadequate, indicated[4] a surprisingly sharp rise in the level of concern for the quality of the environment (Klineberg, 1999). The following section draws on the APCHS. In contrast to the other two surveys, the APCHS has been designed to include the socioeconomic and geographical variables of the residents. Also, the comparative strategy of the APCHS makes it different and more appropriate to the political-economy study of urban air pollution.

The APCHS consisted of a questionnaire presented to 300 households that were located near to as well as at a distance from the city's industrial complex. A *household unit* consisted of members of one household who are a couple or one person without a partner and any of their children, provided these children have never themselves been married and have no children of their own (Office of Population Censuses and Surveys, Central Statistical Office, 1973). The term *child ill health* is used here to indicate those households which reported at least one child suffering from at least one recurrent health disorder. Notwithstanding that most reported symptoms are respiratory, it was decided to employ the more general term ill health to avoid narrowing the effect of air pollution to only one type of illness. Rise in air pollution may also trigger, for example, stomach problems or headaches, as the current study shows. Subsequent sections draw on the data from the questionnaires produced in the APCHS (see Appendix). The local residents in their households, usually the mother figure, provided the information. According to the critical realism methodology of intensive research, clusters of geographical and socioeconomic characteristics classified households. These are low- and high-income households located in the Ship Channel area, with high levels of air pollution and situated near the petrochemical and other industrial facilities, and the Southwest, the control area, far from the industrial sources of toxic emissions. The information collected from local residents represents the variables that are correlated quantitatively and comparatively to narrow down the list of possible biophysical and social factors to those which might have had relevant powers to cause ill health. Bi- and multivariate statistical analyses, with one dependent variable, and regression, have been used to measure and explain common and distinguishing properties of variable relations. The P-values that accompany some associations highlight the significance of the association and are by no means intended to infer causality.

Prevalence of respiratory problems

Most indicators rated very similarly in the Ship Channel and the control area. The APCHS found that shared housing facilities in the poor Southwest study area were poor, that vegetation growing on common/public grounds was scarce, staircases and landings were normally dirty, and sections of the buildings commonly showed signs of vandalism. These findings agree with Urrutia-Rojas's (1988), which show the inferior and precarious living conditions of low-income population in poverty pockets in the Southwest and that unemployment was very high, at 26.6 per cent (this compares to 8.9 per cent in the state of Texas and 7.0 per cent in the US). On the other hand, the high-income area in the Southwest was very prosperous from the point of view of housing, commercial centres, road networks, medical facilities, and so on (see Appendix).

Simple frequencies of aggregated data of reported ill health, but without specifications of house location and income, represent the initial pool of information about the magnitude and main types of disease afflicting children in a city with strong connections to the global economy. Perhaps one of the most important issues to emerge from this analysis was the overall high number of households with reported child ill health. Indeed, 68 per cent of all surveyed households (n = 300) reported at least one child with recurrent illness. Examination of the specific reported health problems that afflicted children offered a further insight, this time on the particular type, and perhaps severity, of widespread ill health in Houston. An overwhelming proportion of the reported health disorders were of respiratory origin (94 per cent) (see Figure 5.6). In terms

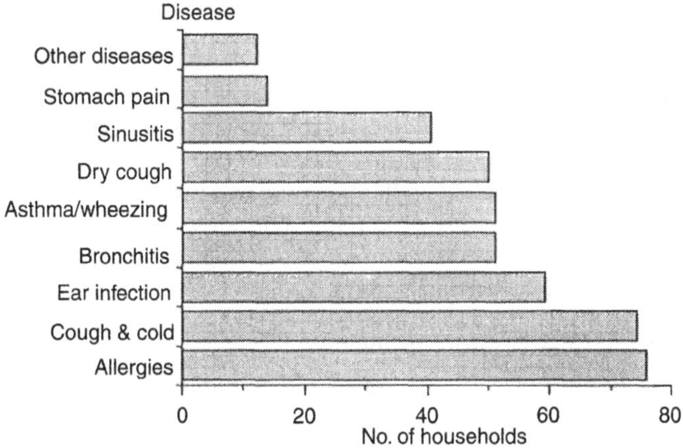

Figure 5.6 Prevalence of reported disease* in the surveyed households
*The total number of reported disease was 453.

of disease, coughs and colds, allergies, ear infections and bronchitis were often reported. A considerable per centage of households reported children with asthma/wheezing. The literature indicates that either poverty, residential location, or wider environmental factors can contribute to such a particular pattern of respiratory disease prevalence (Chapter 2).

The role of income on the incidence of ill health

Although the level of household income has rarely been included in clinical data as a predictor of child ill health, poor people suffer more ill health. Hence, the following sections examine the relationship between six recognized dimensions of inequality that might affect child health. These are: household income, mother's occupational class, access to health care, dampness, number of resident children and parental structure, and the incidence of ill health. Since the social and demographic composition of households in the Ship Channel and the Southwest area is very similar, one rational conclusion would have been that given that children were brought up under these similar household circumstances in the polluted and the less polluted areas, the state of child health ought to be similar. The rest of the chapter deals with this assumption, which, in the light of the findings, proved wrong.

The monthly income of the surveyed households ranged between less than $1000 to more than $3000 (see Figure 5.7) with the same number of households in low- and in high-income categories.

More child ill health was reported among low- (75 per cent; n = 150) than high-income households (61 per cent; n = 150). This finding corroborates the effects of social inequality on child ill health as described in the socioeconomic literature (see Chapter 2). None the less, the gap between the incidence of child ill health in low- and high-income households in the Houston sample population was small (14 per cent) despite the three+fold differential between the lowest- and highest-income group's average monthly income (see Figure 5.8). Graphical representation of this association shows both decrease and also increase of the incidence of child ill health with variations of household income. Both extremes of the relation move in the same direction; that is, negative and positive directions dominate the relationship between child ill health and household income; in other words, a rise in child ill health correlated not only

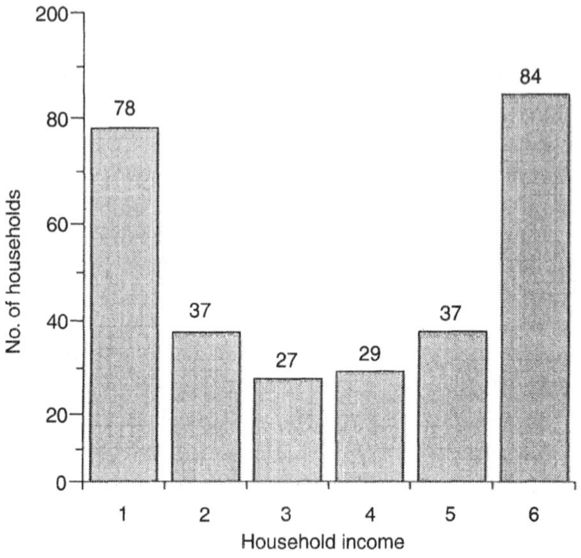

Figure 5.7 Household* average income** of the sample population
*n = 242 households.
**Average monthly income: 1, Less than $1000; 2, between $1000 and $1500; 3, between $1500 and $2000; 4, between $2000 and $2500; 5, between $2500 and $3000; and 6, more than $3000.

with decreasing but also with increasing income. This basic contradiction was a preliminary indication that widespread child ill health was not only, or necessarily, the consequence of poverty in the household. This unusual pattern begs an explanation of the role of further factors in the incidence of child ill health.

The importance of geographical location

The role of household geographical location in the relationship between the extent of child ill health and income is incorporated here. As established in Chapter 2, location is important not just as context but as, perhaps, indicative of wider social processes because the character of the space has been socially constructed, a shortcoming appearing in some social and scientific explanations of ill health.

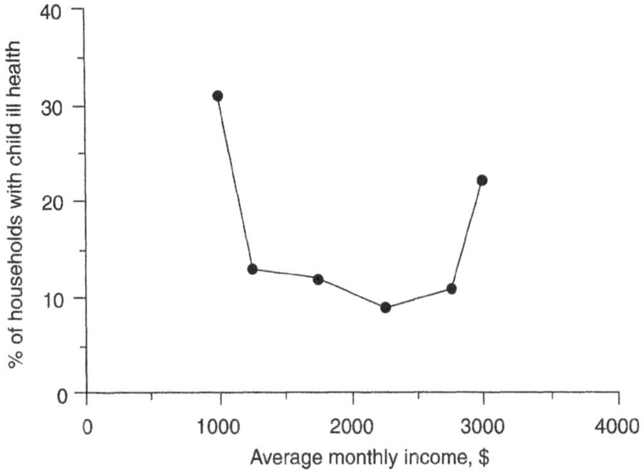

Figure 5.8 Child ill health by household income

Analysis of the role of household location was essential because social, geographical and environmental changes complement each other (see Chapter 2). Significantly, 10 per cent more households (n = 150 households) in the Ship Channel residential area (that is, those near to sources of industrial air pollution) than in the Southwest (n = 150 households; relatively far from sources of industrial air pollution) reported child ill health (73 per cent and 63 per cent respectively) (see Figure 5.9).

Now, an analysis of the geographic distribution of child ill health accounting for the specific different income per household shows that child ill health was similarly and significantly ($P < 0.003$) distributed in lowest-income households in the polluted as well as in the less polluted areas (72 per cent and 77 per cent respectively). However, and perhaps most importantly, the following findings show a strong correlation between the location of the household and the incidence of reported child ill health. While in the households in the control Southwest area the extent of child ill health decreased with rising household income, in the Ship Channel area the incidence remained high with rising household income. Further, child ill health was notorious in all the high-income households, at

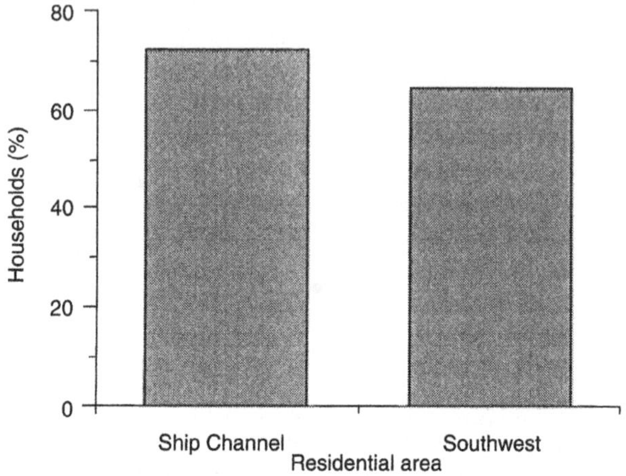

Figure 5.9 Geographic distribution of reported child ill health in Houston

61 per cent. Significantly, while there were overall 10 per cent more households with reported child ill health in the Ship Channel area, the incidence of reported disease was 24 per cent higher in wealthy households in the Ship Channel area than in the rich Southwest (73 per cent and 49 per cent ill health in the Ship Channel and in the Southwest) (see Figure 5.10). Therefore, despite the higher numbers of ill children found among poor households, household income could not act as the exclusive and convincing predictor of the incidence of child ill health.

Evidently, geographic location in relation to industrial sources of air pollution played an important modifying role in the extent of ill health reported by the public in the surveyed households. While it has been established that spatial location *per se* cannot account for the events, it can make a crucial difference to how social and other processes work and to what forms result (as shown in Chapter 2). The next chapters will expand the socioeconomic examination to verify whether there are particular social or demographic conditions present in one area but not in the other that made the spatial factors so relevant for household variation of child ill health. We explore now the ways in which residents recognize local air pollution and what effects it causes to the living areas.

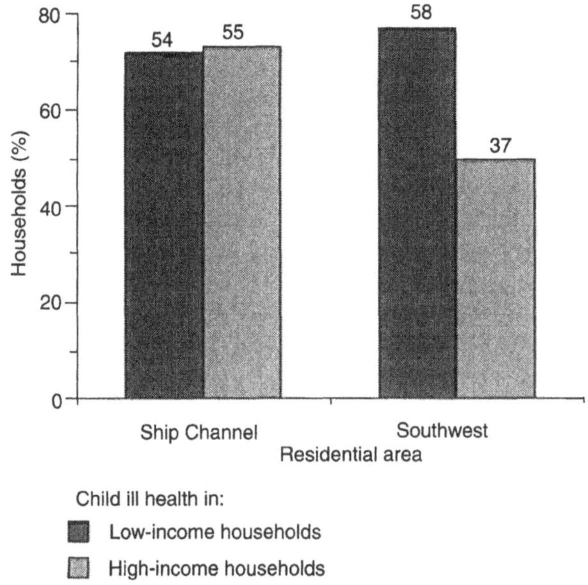

Figure 5.10 Geographic and socioeconomic distribution of reported child ill health in Houston

Odours, government involvement and health risks from pollution

Relying only on the measures of air pollution by the air control agencies poses serious limitations in determining actual health risks, as concluded above. Hence it serves further to justify the elaboration of a database containing the local population's views on air pollution and health risks. In this sense, the authoritative opinion of McMullen, Director of the City of Houston Air Quality Control Board in 1992, is persuasively relevant to our point. He emphasized that: 'A survey of the people who live in the Ship Channel can be much more reliable than measures of emissions' (McMullen, personal communication, 1992).

The issue of air pollution has preoccupied the population of Houston and has also gained a certain degree of government attention due to complaints from the residents (TACB, 1993b). In

December 1992, the TACB appointed a 'Nuisance Odors Task Force' (NOTF). Apparently, one of the main reasons for the creation of the NOTF was the 'public's desire for action' (ibid., p. 4). The new department in the TACB intended to formulate a basis for new policy on 'what members agree is a difficult, subjective area of regulation'. The main plans of the force were to study new advances in measuring odorous air pollution as well as methods to assess the severity of such odours. Despite the evident importance of the task of this agency, it was only empowered with managerial and technological tools – but had no law enforcement powers. Because air pollution was interpreted only as a nuisance, rather than a hazardous condition which may expose the population to great health risks, the force's scope for bringing about changes was narrow. Indeed, tracking nuisance odours to a specific source seemed a difficult task for the NOTF:

> Odors come and go with the wind and come from various sources, so at times it's hard to track down what the source is. We really respond quickly now as a local pollution control program, probably within minutes sometimes. But often we have not been able to confirm, to the satisfaction of the complainer, that a nuisance actually existed.
>
> (p. 4).

Most complaints were triggered by industrial activity (for example, smell from the refineries, spills of transported toxic materials, explosions in chemical plants), although other reasons for complaint may cite bakeries, garbage, neighbours and so forth. In the household survey, residents have clearly acknowledged the presence of air pollution in their local environments and have designated it as a real nuisance – even as a threat.

Altogether, 79 per cent (n = 300) of the sample population reported air pollution (only 2 per cent were not able to give an opinion and 17 per cent reported that they had not noticed air nuisances): 'We always have the smell of the refineries. You know the smell from the Ship Channel' (quoted from a resident in a low-income household in the Ship Channel area).

Air pollution may have been even worse than reported. Some airborne toxic substances are either odourless, or their typical odour cannot be associated with noxious materials which may have impeded their recognition. Such is the case of ozone formed in the atmosphere from the combination of nitrogen oxides, additional toxic components and sunlight. The odour of ozone resembles, in fact, the 'smell of the air after a summer storm' (McMullen, personal communication, 1992; see Appendix). Therefore, in spite of the fact that ozone concentrations may reach high levels in both the Ship Channel and the Southwest areas, the residents can hardly distinguish it. On the other hand, sulphur dioxide, carbon monoxide and particulate matters that reach high levels of concentration particularly in the Ship Channel can be more easily singled out. Residents may therefore more readily recognize this pollution but may easily miss the presence of atmospheric ozone. In this way, some air pollution may remain unidentified and, therefore, will not have been adequately reported in the household survey.

The disguising effect of other types of very localized air pollution may occlude the presence of chemical environmental pollution. In the household survey, stench from communal dustbins and also from sewage was the commonest nuisance reported apart from air pollution. Only 16 per cent of the surveyed households reported air nuisances other than chemical. In particular, garbage odour was very disturbing for 13 per cent of the interviewed population, with sewage being the source of unpleasant smells for 2 per cent of the respondents. Out of all low-income households in the less polluted Southwest, 37 per cent (n = 75) of the respondents complained of unbearable 'dustbin odours'. Also in the Southwest, but in high-income households, residents reported odours from sewage (8 per cent). They attributed this stench to the Braes Bayou, a nearby polluted stream that cuts across the city. In the polluted area, only a few households (4 per cent) reported odours from garbage and sewage.

Indeed, air pollution was the most frequently reported preoccupation in the polluted study area, the Ship Channel (49 per cent; n = 150; other preoccupations were personal safety – fear of robbery, crime and drugs, water pollution, quality of local schools, and health care). In the Southwest, which is less polluted, the worst problems reported in order of frequency of citation were personal safety followed by water pollution, and air pollution came only third (15 per cent; n = 150). The APCHS revealed that more high-income – 62 per cent in the Ship Channel area and 78 per cent in the Southwest – than low-income households showed concern over air pollution. None the less, reporting of air pollution as the worst local problem was remarkable in low-income households, at 38 per cent in the Ship Channel and 22 per cent in the Southwest. This is a notably high per centage for low-income households because preoccupations such as environmental concerns would be expected to be overshadowed by health care or personal safety. The degree of reporting indicated that the level of environmental pollution to which residents were exposed in Houston in general, and in the east side of the city in particular, must have been considerable.

At the outset, a strong statistical association was found between reported air quality in the surveyed households and the spatial location of the household (P = 0.000). This indicates that geographical location might be influential in the extent of reported air pollution. While the overall per centage of households that reported local air pollution was remarkably high, air pollution in the Ship Channel area was twice as heavily reported (85 per cent; n = 150) as in the Southwest (42 per cent; n = 150). Also the degree of reported severity of air pollution varied from area to area (see Figure 5.11).

The extent to which the air was pungent, and the frequency with which it became polluted, determined the severity of air conditions. In the polluted area, severe and moderate air pollution was reported (34 per cent and 49 per cent respectively). In the less polluted area, considerably fewer households reported severe air pollution (5 per cent) compared to clean air (42 per cent). While some households also reported other types of air nuisance (for example, odour from dustbins, sewage), clean local air was reported by only 15 per cent in the Ship Channel area but by 58 per cent in the Southwest.

In order to compare and integrate the extent of air pollution reported by the residents with that measured by the governmental agencies, the household survey and the 1990 reports from five monitoring sites were compared. The government sites widely correspond to the geographical clusters selected for the household survey. Although not all the US standard pollutants had been sampled regularly in the Houston area during 1990, complete records were available for ozone and sulphur dioxide (TACB, 1993a; City of Houston Department of Health and Human Services, 1990). Fortunately, these alone certainly reflect the distribution of air pollution in the city (see Table 5.3).

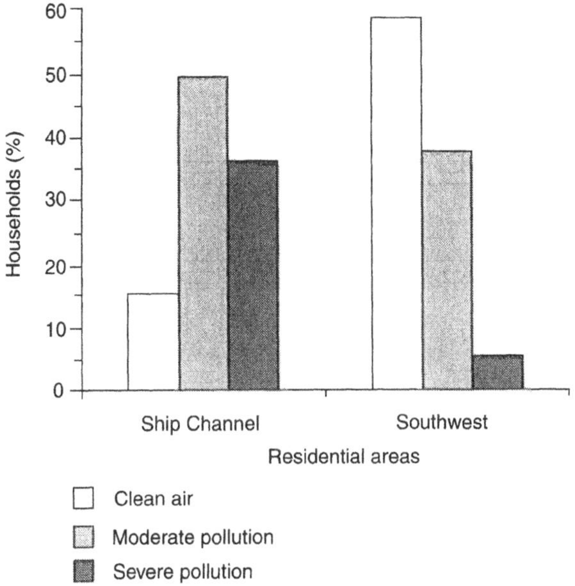

Figure 5.11 Reported air quality in the residential areas

One interesting point to be noted here is that remarkably more sulphur dioxide was measured in the Ship Channel area compared to the Southwest. Ozone, however, was very high in both areas. It should be remembered that ozone spreads out and that concentrations of ozone found in the Southwest side of the city had partly been displaced from the petrochemical plants, as the ten-year study by the City of Houston has demonstrated (see Chapter 4). The next noteworthy issue is that the extent of air pollution which interviewed residents reported seems to be fully backed up by the measurements originated in monitoring machines. This represents a very important finding for agreement between monitored and residents' reported air pollution, which provides a scientifically verified foundation to any conclusion drawn from the information given by the local residents.

Hence, a fuller picture is obtained. As we shall see, local people's reporting, in contrast to monitoring, highlights the gravity of the

Table 5.3 Comparison between measures of contaminants in Houston monitoring stations and residents' reporting of local air pollution*

	NAAQS					
Residential areas and monitoring stations	*Machine monitored pollution*					*Resident's report of local pollution (% of households)*
	O_3	*CO*	SO_2	NO_2	*PM-10*	
Ship Channel						**Ship Channel**
East						
Clinton Station	0.23	5.2	0.008	0.024	44.7	
Crawford	0.22	7.9	0.006	0.029	30.3	
Southeast						
Monroe	0.23	N/S**	0.003	N/S	28.0	
Northeast						
N. Wayside	0.23	N/S	0.006	N/S	N/S	85% of total households
Southwest						**Southwest**
Croquet	0.22	N/S	0.002	N/S	N/S	42% of total households

* Lead is not included in the table because levels have not been exceeded in Houston.

** N/S: not sampled.

problem by clearly describing direct effects of contamination on their children and sharply addressing additional dimensions of air degradation, like stench and visibility, within the context of household location in relation to sources of industrial emission. Importantly, whilst all possibilities of examining aggregated government data had already been exhausted by the current research, there is considerable scope for further exploration of the subject using information from the participatory survey, as the results of the current enquiry demonstrate. This way of acquiring knowledge is significant because it can provide the necessary elements to substantiate future economic and environmental policies.

Geographical distribution of child ill health

The geographical distribution of reported chronic child ill-health problems reflects the impact of air pollution on the residential area.

A distinctive spatial concentration of most recurrent child illness reported was found in the polluted area. Here, the prevalence of some conditions such as wheezy chest/asthma, stomach upset and coughs and colds was remarkably higher than in the less polluted residential area. The incidence of sinusitis and bronchitis was also higher in the Ship Channel area than in the Southwest (see Figure 5.12). Therefore, a clear cluster of child ill health, respiratory in particular, was identified in the residential area near the industrial sources of air pollution. sources of air pollution.

It should be stressed that reported child ill health was remarkably high also in households far from the sources of industrial emission, a factor which could relate to the large concentration of reported child illness in low-income households. Although the prevalence of illness in low-income households was not very different in the Ship Channel (72 per cent) from the Southwest (78 per cent), poverty seems to be a more appropriate factor for ill health in the Southwest (see Urrutia Rojas, 1988). Thus, it is likely that economic and social disadvantages were major propagators of illness. However, this assumption does not invalidate the possibility that the general level of background air pollution may also have been a trigger for ill

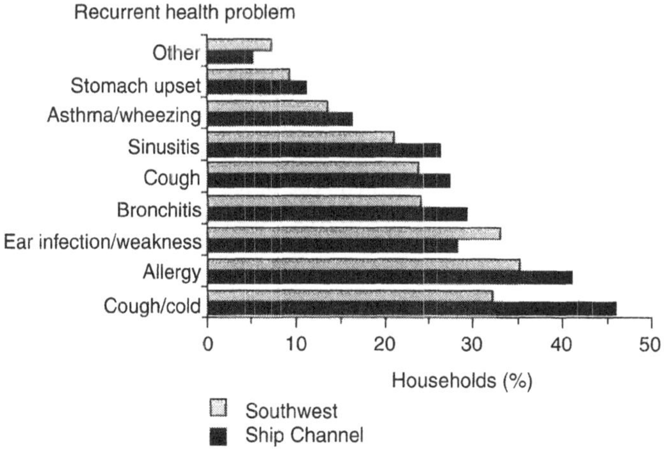

Figure 5.12 Recurrent child diseases by geographic areas

health in low-income households in the Southwest. It should be remembered that although these households are located far from the sources of pollution, ozone pollution is also very high in this part of the city. None the less, it is unlikely that it will be possible to assess the separate impact of either factor due to the strong influence of poverty in this area.

Particularly interesting is the prevalence of reported respiratory child illness in high-income households, because reporting here is less prone to the confounding effects of social disadvantage. In the Ship Channel area, 23 per cent more high-income households than in the Southwest area reported child ill health (72 per cent and 49 per cent respectively; n = 75 each). Out of all reported cases of asthma and wheezy chest in high-income households (n = 33), 70 per cent were found in the polluted area. Furthermore, 76 per cent of all bronchitis cases (n = 2), 80 per cent of children with stomach problems (n = 5), and 62 per cent of all children with chronic cough (n = 32) were found in the polluted area. More high-income households in the polluted than in the less polluted area reported incidence of allergy (56 per cent; n = 43) and sinusitis (56 per cent; n = 26); these were also very high in the less polluted area (44 per cent and 44 per cent respectively). Only ear infection prevailed in the less polluted location to a greater extent than in the Ship Channel area.

Health risks from local air pollution

Within the perspective of highlighting conditions that help to identify the adverse impact on people of air contaminants, the residents' survey examined whether rising air pollution caused immediate health response and, if so, which type of reactions prevailed. It was thought that identification of health changes during and immediately after a rise in pollution levels, had they occurred, would provide robust proof that particular social processes were activated through biophysical mechanisms. Such a study adds to the existing epidemiological type of data. For example, epidemiological studies have been done to compare daily levels of particular air pollutants with health reaction (Rutishauser et al., 1990; Schwartz and Marcus, 1990; Stock et al., 1988). These have used badges that are put on individuals to measure exposure, or employed monitoring reports and spirometric tests to assess breathing changes. The problems with these techniques are, first, that health changes are registered at a fixed time; and they measure respiratory symptoms only, leaving out other health problems which may also be triggered by air pollution, for example, headaches and eye soreness. Neither could an epidemiological assessment alone take notice of passing, yet important, comments by the residents, such as the following: I haven't noticed yet any change in my children's health but my two dogs get sick and weak every time we have the smell from the Channel' (quotation from a mother in a low-income household in the polluted Ship Channel area, Houston, 1990).

The case study was essential in procuring this information on whether local air pollution has had an immediate effect on health. This was crucial for identifying the interaction between natural mechanisms and social processes as defined in the causal model. The health effects of rising air pollution were examined in the sample households that had been classified by similar socioeconomic status, and geographic location, with the purpose of explaining the distribution and the causes of the incidence of child ill health. In this sense, the intensive research approach to the households in the survey and the comparative analysis of the data convincingly complemented the analysis of government reports of air pollution and of public health status in Houston.

A fairly large number of interviewed households (41 per cent; n = 300) reported that any worsening of air pollution affected the health of their children. It is very likely that the reporting of health changes following rising air pollution would have been higher had it not been that a considerable percentage of households (24 per cent) did not pay enough attention to rising air pollution and health changes. One-third (34 per cent) of the sample households reported that they did not notice visible health changes when air pollution rose. The most often reported health responses to air pollution were allergy, headache, shortness of breath, and weakness, together with eye soreness (see Table 5.4).

The risk from breathing polluted air evidently posed a health threat to Houston residents. It is more evidently so for residents, and for children in particular, who live in the proximity of the oil and petrochemical complexes and port facilities along the Ship Channel. The distribution of reported health responses to rising air

Table 5.4 Distribution of health changes according to spatial location of the household following worsening air pollution

Health change	*Total symptoms (100%)*	*Ship Channel polluted area (64%)*	*Southwest less polluted area (36%)*
Allergy	46	54%	46%
Headache	27	89%	11%
Shortness of breath	20	65%	35%
Weakness, eye soreness	18	72%	28%
Cough	10	30%	70%
Stomach upset	2	50%	50%

pollution (n = 123) followed a clear spatial pattern. The majority (64 per cent) of reported reactions originated in the polluted area.

Focusing on the polluted area only, both low- and high-income households showed considerable awareness and concern about the health effects of deteriorating conditions of local air quality. None the less, sensitivity towards air pollution and child health seemed to be more developed in high- than in low-income households (60 per cent and 40 per cent respectively). Mothers in low-income households in the polluted area tended also to attribute child-health difficulties to a variety of reasons related to social inequality rather than to environmental conditions. While the relation sounds plausible, and the survey has shown significant correlation between poor-income indicators and ill health, a significant fraction of child ill health in low-income households, particularly in the polluted area, might also have been triggered by contaminated air.

In high-income households, on the other hand, reported health changes as a result of more air pollution were remarkably higher in the polluted than in the less polluted area. For example, every time that pollution levels rose in the Ship Channel area over a quarter of children had headaches (26 per cent) but only a fraction (4 per cent) reported this in the Southwest; well over a third (39 per cent) developed weakness, uneasiness and eye soreness in the Ship Channel area as did 17 per cent in the Southwest; 33 per cent reported allergies in the polluted and 17 per cent in the less polluted area and, finally, a rise in air pollution triggered coughs in 20 per cent in the Ship Channel and none in the Southwest.

Length of stay in polluted areas

It was assumed that longer residence in the air-polluted area would correspond with more reported child ill health due to prolonged exposure to toxic air. While there are hardly any studies which either measure exposure to air pollution over an extended period (see Girt, 1972; Pope, 1989), examine the actual health changes, or estimate the cumulative risks of contracting ill health as a result of prolonged exposure, there is some earlier evidence of suspected long-term effects of prolonged exposure to air pollutants (Bland et al., 1974; Colley et al., 1973; see Holland et al., 1978).

Residence at the same address or geographic area was classified in periods of less than one year, between one and three years, between three and five years, and more than five years. Residents in the polluted area had resided for considerably longer at the same address than residents in the less polluted area (see Chapter 7). It was found that in the Ship Channel the longer the family lived in the area, the higher the number of reported households with child ill health. In the less polluted area, in contrast, highest incidence of child ill health was correlated to shorter residence in the Southwest (see Table 5.5) which was more often found in low-income households (21 per cent) stricken by poverty and lack of medical services (see Figure 5.13).

No clear pattern, however, has emerged between the length of exposure and child ill health in low-income households in the polluted area. It is presumed that poverty may play a confounding and compounding role in the relationship between air pollution and child ill health. In high-income households, on the other hand, reported child ill health persistently increased, as exposure was more prolonged. Thus the findings strongly support the argument

Table 5.5 Distribution of child ill health according to spatial location of the household and length of exposure

Reported child ill health	*Less than 1 year*	*Between 1 and 3 years*	*Between 3 and 5 years*	*More than 5 years*
Ship Channel	19%	31%	21%	30%
Southwest	33%	34%	22%	12%

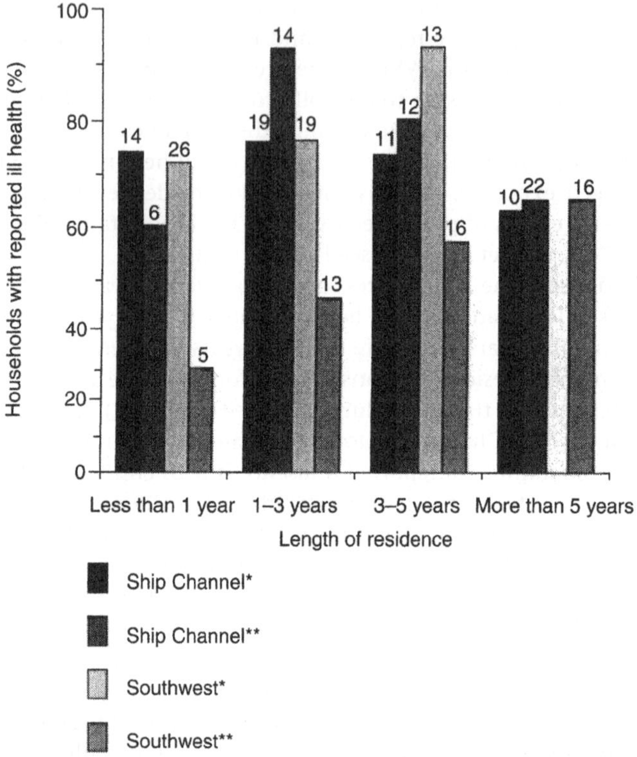

Figure 5.13 Social and spatial variation of reported child ill health by length of residence
*Low-income households (no household in the Southwest for the more than 5 years category); ** high-income households.

that prolonged exposure to air pollution contributed to the deterioration of child health in the Ship Channel area.

In summary, unpleasant and dangerous living conditions imposed by continual exposure to air pollution from the petrochemical plants is a feature of the Ship Channel residential areas. Health changes in children have been identified. Although contaminated air is also found in other parts of the city, e.g., the Southwest, the degree of severity is undoubtedly greatest in the area close to industrial sources of emission. One local resident vividly, and sharply, expressed grave concern for her children's well-being:

> I do not want my children to grow up in this neighborhood, although it is good and pretty. The refineries pollute the air. My neighbors are mostly widows; their husbands have all died from cancer. I do not want my children to live here.
>
> (Quotation from a parent in a high-income household in the Ship Channel, Houston, 1990)

Conclusion

The central question which has driven this book is the identification of the underlying social forces that promote persistent environmental degradation in world cities, and, hence, consequent ill health. In addition to the fact that certain pollutants are known to be particularly damaging to health, lower levels of air pollution also seem to represent a considerable problem. This was our conclusion in the epidemiological review of Chapter 2, but has now also been attested by the moderate incidence of reported child ill health in the less polluted Houston area. The US pollutant safety standards seem inadequate to protect Houston's residents' health both in the short and in the long term. This chapter has shown that air pollution has been a constant problem in Houston since records were first kept in the 1970s. However, the quality of the urban air had been a preoccupation well before this decade, as signalled in the previous chapter. The magnitude and quality of pollution have none the less altered. Decades of economic growth and globalization have resulted in differentiated geographic configurations, with varied levels of pollution within the city as seen in relation to the Ship Channel and the Southwest area which vividly represent the uneven environmental outcomes.

Pungent odours originating in the oil refineries have been a chronic preoccupation of residents living in Houston, one of the most prosperous residential and industrial areas in the US. Possibly, pungent chemical odours have awakened residents to some of the risks involved in unregulated growth. However, there have been further consequences to constant, and at times severe, air contamination. These appear in the form of ill health among the children of this world city, particularly for those living in the east side. On the basis of the case study, we learnt of the strong correspondence between machine-monitored air pollution and the level of disturbance reported by local residents.

The degradation of the natural environment in and around Houston, particularly in the Ship Channel area, is such that living conditions for substantial numbers of the population are too often intolerable, their health and that of their children being sacrificed to the capitalist ethos. Dangerous concentrations of ozone and sulphur dioxide, monitored since the early 1970s, have evidently become accepted parts of Houston's everyday life. Uneven levels of contamination across the city cannot be dissociated from historical trends that have favored industrialization as the necessary way for attaining better living conditions. The conjunction of Houston's regional growth and global links has contrived unique spatial local configurations whereby industry and housing share adjacent territory. The prevalence of large amounts of industrial toxic emissions and their interaction with atmospheric conditions, which favour displacement, and the formation of new pollutants, mean that the entire city has been dangerously exposed to various levels of industrial pollution. Today, this represents a common aspect of many large cities, in the developed and developing worlds alike.

The combined effect of social structures and biophysical mechanisms that produce air pollution and ill health has been most prominent in the Ship Channel area. Here, highly developed petrochemical establishments and other industries have grown continuously to the point where the city has won international recognition for its remarkable economic achievements but at the same time is confronted by severe ecological consequences. The relationship between industrial air pollution and child ill health is essentially a reflection of the effects of the more general process of wealth creation and the resulting environmental degradation. The contribution of this chapter lies in the fact that the particular type of information obtained on child ill health and on transformations of local air pollution across the city can bridge the existing gaps between historical developments and monitored contamination. This book recognizes that the current events of local pollution are a window to detect the power of certain social structures to transform nature. According to the multi-faceted model of Chapter 3, this study has so far applied a social structural account of past and present development events, and discussed participatory information from local populations. The existence of air pollution in modern Western cities in the first place depends upon the type and extent of economic activity, and upon the institutional efforts and government's commitments to protect the environment and public health. Climatic conditions, particular chemical reactions and topographic characteristics are also involved in air pollution.

It has been established that the level of child ill health in Houston is well above what might be expected. The cause of this elevated incidence of ill health is not attributable only to conditions of household poverty. Distinguishing between the effect of household income on the extent of child ill health in practice, and the role of spatial location of the household in relation to industrial sources of air pollution, this chapter has demonstrated how key a factor location is for intensifying health risk. Yet such determination only indicates that there are wider social processes occurring over space. The longer children had resided in the same area, that is, in the Ship Channel area, the more residents reported the ill health. The length of exposure to industrial air pollution was a definitive indication of the consequences of continually breathing some of the externalities of industrial development. Short-term, as opposed to chronic, reported symptoms – and those which were not only respiratory, for example, headache, allergies and eye soreness – were again significantly associated with reported raised levels of air pollution, a phenomenon more often registered in the Ship Channel area. This information was made available because residents were sampled according to the characteristics of the residential areas, that is, within their 'causal' and structural contexts. This is a procedure that draws on the critical realist methodology and illustrates the usefulness of comparative study within an extensive and intensive research design. The public's information under this design has the potential to capture the 'combination' effect of structural and spatial factors and the uniqueness of their own realities.

A sudden change in the levels of pollutants appears to have acted as an environmental trigger for the reported incidence of short-term ill health symptoms, particularly among children who lived in the proximity of the Ship Channel area. Furthermore, prolonged residence in air-polluted areas has been associated with reported chronic health complaints. The two banks of the Houston Ship Channel (the strait that extends from the Gulf of Mexico, crossing the so-called ring-road 610 Loop) penetrate far into the city, displaying densely industrially developed land.

Whilst the survey has found a very high incidence of child ill health in the sample households in general, the comparative analysis has uncovered that, as expected, the incidence of illness among poor children is always very high, indicating that social inequality was an important factor. However, one of the most revealing findings of the participatory database is that the incidence of ill health in all high-income households was also unexpectedly high. Considerably more than expected child ill health had been reported in high-income households in the Ship Channel area, that is, as compared to the Southwest, in the most air-polluted area. This point will be further explored in Chapters 6 and 7.

This chapter has revealed important facts about the city, such as increased air-pollution levels and clusters of unrecorded child ill health. The annual measures of air pollution in the east and Southwest sites clearly indicated higher pollution in the Ship Channel area, notwithstanding remarkable ozone concentrations in the Southwest. On the other hand, the degree of air pollution reported by residents also pointed to variations in the degree of pollution. The technical information alone, however, was not sufficient to reflect the health impact on the population, if any.

Epidemiological studies had firmly determined a relation between urban pollution and ill health. It was necessary to assess this same correlation for groups of people who are exposed to pollutants, and to other factors. Yet the ability of the population to deal with health risks from industrial pollution seems to vary according to where and how residents live. While the current chapter focused on the first question, the next one will look at the effects of the second interrogation. The crucial theme to discuss now is whether the priorities of the institutions of health care and environmental protection match the need to protect the environment and the population. Chapter 6 will deal with the first issue.

Notes

1. The standards are further categorized for different averaging times. Long-term standards specify an annual or quarterly mean that may not be exceeded; short-term standards specify upper limit values for 1-, 3-, 8- or 24-hour average time spans. Except for the pollutants ozone and PM-10 (particulate matter), the short-term standards are not to be exceeded more than once per year (US EPA, 1991, p. 2–1).
2. For example, in the summer of 1988, which was exceptionally hot, particularly in the east and along the north US coast, extreme rises in ozone levels were registered. In that year, 101 areas in the country failed to meet the 0.12 ppm ozone standard.
3. Lead is the only major pollutant not included in the index because it does not have a short-term NAAQS, a Federal Episode Criteria or a Significant Harm Level.
4. The *Houston Area Survey* included environmental issues among other city concerns and has been carried out annually since 1982 by the Rice University at Houston under the direction of Stephen Klineberg.

6
Social Inequality and Health Risks

Introduction

Chapter 5 has indicated that although environmental regulations put in place since the 1950s and the 1970s may have ameliorated some of the degrading effects of economic activity, they have mostly failed to reverse the generally rising trend of pollution and maintain harmless levels of air pollution in many cities. The benefits arising from improvements in new combustion technologies have been outstripped by rising output resulting from international economic growth (see Chapter 4). The current chapter argues that relying on improved access to health care may provide temporary alleviation for low-income population to cope with the effects of pollution. It cannot, however, guarantee any lasting solution to the wide and unknown consequences of exposure to air contaminants experienced by so many residents, particularly in large cities of the developed world.

Previous chapters have discussed how the underlying causes of rising air pollution in cities, and of consequent related ill health, have generally been socially induced. The study of the main societal forces which have instigated transformations in the quality of the air and health have, however, often been masked by an exclusive medical and ecological focus. Retrieving past economic and political developments has been important to appreciate the continuity between past and present concentrations of pollutants, as analysed in Chapter 4. The case of Houston has shown how unregulated economic growth has generated all types of ambient contamination.

As stated in the multidisciplinary model, the study of current local institutions is also necessary to establish the relative effect of these institutions on people's health and pollution in world cities. Chapter 6 addresses the role of the US medical care system and individual socioeconomic circumstances on their health status. This enquiry is in line with the thought, developed in Chapter 2, that ill health is the likely outcome of an inequitable health care system, poor socioeconomic conditions in the household, the system of production, and exposure to industrial externalities. Chapter 7 will further develop the topos of how environmental regulation relates to health.

Chapter 5 has revealed that in Houston, moderate to high levels of pollution have prevailed at least since official records started in the 1970s. The NAAQSs have been constantly breached, and ozone levels are among the highest in the US. The PSI > 100, designed to protect health, has been repeatedly exceeded during the period between 1984 and 1993. The data drawn from local residents indicate significant incidence of ill health in two areas of the city, but with higher prevalence on the east side. A remarkably high number of children was discovered to suffer from pollution-related illness in high-income households in the Ship Channel area. Chapter 2 stated that poverty represents a main cause of ill health, a premise that is partly challenged in the light of the case study findings. Chapter 6 assesses the effects of the particularly selected sociological indicators for inequality in order also to explain the unusual, though not abnormal, findings of high levels of illness in high-income households. This step is necessary to elicit the effect of social inequalities in comparison to those of air pollution.

As in most economically developed countries, the US health care sector in Houston is highly proficient. Houston houses some of the largest and most important medical centres in the world, and boasts advanced research and cancer treatment expertise. Numerous facilities – among them 50 hospitals, 15 651 beds, many clinics and convalescent homes and virtually every medical specialty – are represented (Greater Houston Partnership, 1995/96). It would be expected, therefore, that sufficient health care mechanisms would be provided to protect the public's health. However, this has not been the case and this chapter furnishes evidence of how the contradiction between nature and society can also be reflected in the wealth and poor health of the people. Medical facilities which care for the uninsured and the poor do exist, but are few and inadequate. A clear contradiction exists between the glamour of the world city and an eminent medical industry, on the one hand, and the private character of medicine that controls most of it, on the other. In Houston, as in any other US city, access to health services is limited. In fact, it is questionable whether the health of the regional population, rather than the growth of the 'medical industry', has been the main priority of the ruling government.

The chapter discusses the medical care system, the general standard of public health in the city and, finally, focuses on the database. There then follows assessment of variation of the health of children in response to significant socioeconomic inequalities and the geographical location of their households.

Social exclusion and access to health care

In 1990, the G-7 Economic Summit was held in Houston. Sorelle (1990b) stressed then that the US health care system costs more and covers a smaller percentage of the population than systems of any of the other six economic powers. Infant mortality was higher and life expectancy lower in the US than in countries with similar, or even lower, GNPs. Canada, France, West Germany and Japan finance health care through a form of national health insurance. The UK and Italy both maintain publicly provided national health services. Despite the fact that per capita health care expenditure in Italy was two and half times less than in the US, the infant mortality rate was twice as high in the US (Sorelle, 1990b). Moreover, infant mortality in Germany and in Japan was similar to that in the US in spite of the fact that the US expended twice as much or more on per capita health care than the other two countries (see Table 6.1). The portion of the GNP invested in all health expenditures in the US was only 1.2 per cent (Dougherty, C.J., 1988). In comparison, in France and Canada health expenditure was remarkably higher (8.6 per cent) than it was in West Germany, Italy, Japan and the UK.

In general, in the US, health care, and the health insurance on which it often depends, are distributed in the same way as most goods and services – largely on a pay-as-you-go basis. This marketplace mode of distribution is by its nature bound to work against the interests of those least well off, the disabled and single-parent

Table 6.1 Comparison of health care resources of the major developed nations

	US	*Canada*	*France*	*W.G.**	*Italy*	*Japan*	*UK*	*Mean*
% of GNP used for health	1.2	8.6	8.6	8.2	6.9	6.8	6.1	8.1
Per capita health care expenditures, $	2051	1483	1105	1093	841	915	758	1178
Population uninsured (million)	35	0	0	0	0	0	0	0
% of population eligible	43	100	99	92	100	100	100	91
Infant mortality per 1 000	10.0	7.9	8.3	9.6	5	9.1	8.2	
Physician's income as a multiple of that of average worker	5.4	3.7	2.4	4.3	N/A	3.9	2.4	3.7

* West Germany

Source: adapted from Health Data File, Organization for Economic Cooperation and Development, 1989, in Sorelle (1990b).

units. It has been apparent that Blacks and other ethnic minorities, the poor, those with low incomes, and the less educated have benefited substantially less from the health care system than have other Americans (Dougherty, 1988). In general, children as a group have been less well covered by private health insurance and in addition they, or their parents, pay a higher proportion of medical costs out of pocket than any other age group (Miller et al., 1985). It remains the case, argues Edgar et al. (1989), that while access to medical services by the poor has improved in the US since the national public health programmes of the 1960s, those services are inadequate, time-consuming to reach, hospital-based, reduced in scope, episodic and disease-oriented as opposed to the range of services offered by the private medical sector. The usual provider of medical care for most people in the US is a privately practising physician, whose fees are financed in various ways: by private health insurance, by government-supported public health insurance (Medicare and Medicaid for the eligible poor)[1] or by out-of-pocket payments. Unlike that in comparable nations, US health insurance is largely focused on employer-provided plans. Commercial health insurance is thus linked in general to employment. But employment does not guarantee health insurance because many self-employed, temporary and part-time workers, and those working in small businesses and for low wages, are often uninsured or underinsured.

A major reason for deficient health achievements is that health insurance operates as a reimbursement mechanism, rather than as a care system.[2] Other reasons include eligibility criteria barriers for patients, barriers for providers and limitations in the services provided under this system (Children's Defense Fund, 1990). It has been estimated that millions of poor Americans live in areas with a shortage of physicians and in many counties have no available practitioners who participate in Medicaid. The lack of physicians serving the inner-city population is equally notorious (Dougherty, C.J., 1988).

In summary, the root causes of the contradiction between a large expenditure on health but poor health achievements in the US overall can be traced to the way in which medical care is organized. The state of public health in Houston, which if far from excellent, must be framed within the health achievements of US society as a whole – and these have been shown to lag (sometimes well) behind those of other countries in the advanced world.

In Houston, indigent health care is a continuing crisis, but it would take little for it to become a major crisis affecting everybody, not just the poor (Dr Joe Rubio, in Sorelle, 1990a). The bureaucratic limitations of the public health eligibility process result in more than half of all applications being rejected, mostly for incorrectly filled-out forms. In Houston, over 50 per cent of people living below the poverty line do not have any type of health insurance, including national medical insurance. Fewer than half of the children eligible for nutritional support (available to women, infants and children) are actually served by the programme. From 1986 to 1989, the overall Personal Health Services Budget at the City of Houston

Health and Human Services Department fell by 15 per cent and the budget for women's and children's health fell 80 per cent (Children at Risk Committee, 1990). Low-income residents are mainly assisted through public health agencies run by the City of Houston Health and Human Services Department authorities, hospitals, and charitable clinics run by volunteers (like the Clínica María, see Appendix).

For example, for the 750 000 people in Houston MSA eligible for public medical care at the public hospital district in the late 1980s, there were only 11, non-private and run by the city's government, community Health Service Area Clinics and 1035 hospital beds. Moreover, there was no public clinic in the Southwest of Houston. A large concentration of immigrants, particularly from Central and South America, and many of the under- or unemployed are found here and fewer persons have a regular source of health care there compared to the Hispanic population in general and to the US population overall (Urrutia-Rojas, 1988). The only alternative source of health care for most of the population in the poor Southwest is a private physician, which means that a large proportion of an already low annual income must be spent on health care. This means that if a person does not have the money, he or she will be unable to receive medical care even when such care is imperative (ibid.). One half of all the residents' interviewed in the APCHS (n = 150) in low-income households were found in this particular sector of the poor Southwest.

Access to child care in the public sector could not be assured in Houston because there was no organized system of health service delivery or health records. Many private physicians, public agencies and charitable/community agencies provide services, but there is little contact between them (Moyer, personal communication, 1992; see Appendix). To cite one example, the schools require proof of immunization, but access to immunization records is readily available only to a small number of children enrolled in welfare programmes such as Well Child Care. Children who have been immunized at community sites have access to their records only if they can recall the exact day and location of the immunization received. Furthermore, there is no tracking and recall system to ensure that children receive timely immunizations. Each welfare agency in Houston sets different financial eligibility criteria. The combined effect of these uncoordinated operations is that the patient must invest hours completing repetitious paperwork to obtain public services. Furthermore, Houston public transport is inadequate and often inconvenient. It sometimes involves travelling downtown and changing buses in turn, to reach the neighbourhood clinic. Moreover, obtaining vaccinations for babies may become extremely difficult if mothers need to travel carrying one or two toddlers (Hardikar, personal communication, 1992).

There is also fragmentation of preventive and treatment services in the Houston Public Health Agencies. A parent has to travel from site to site, usually from one part of town to another, to access health care from one of the three agencies. Moreover, these agencies usually have different eligibility criteria, fee schedules and residency requirements. Also long waiting times are a chronic problem in the public health care system in Houston, a feature which suggests that the capacity to serve has been overwhelmed by the demand for care. These children wait for treatment not in the plush surroundings of a private physician's office, but in the crowded waiting area of a public health clinic – the Martin Luther King Health Center in the southeast of Houston:

> Every weekday, children and adults line up from opening to closing time to see doctors for minor and major aches, X-rays, and various other needs. The MLK center and the other county and city health clinics are the primary health care providers for the poor. But physicians say the badly needed medical care is not always easily accessible.
>
> (Seay, 1990, p. 4)

Charity clinics fill some of the gaps in the US medical care system. The Clínica María, in the Southwest, was operated entirely by volunteers. In contrast to the usual rules of eligibility stated by the government, there patients do not need to fulfil stringent requirements in order to obtain free health care (Ada Montalvo, personal communication, 1990). The clinic functions in the precincts of a private house whose owners actively promote health care for the poor, in particular, for the *indocumentados*, that is, usually Spanish-speaking immigrants who have not legalized their residence in the US. The staff includes volunteer physicians, medical students and administrators. Medicines, which are donated by pharmaceutical companies, are provided free of charge. Patients obtain primary care and children can obtain immunization that may have been denied to them somewhere else in the national provision. None the less, the extent of the health care provided is necessarily very limited; despite the fact that the Clínica María is heavily used, it is open only a few hours a day, its premises are small, facilities insufficient, and patients must wait for hours before being seen.

In summary, it has been shown that access to health care in Houston is very limited, despite world recognition for its excellent medical facilities. An ethos of for-profit medicine and a bureaucratic process of eligibility to obtain national supported health insurance have hampered the delivery of health care.

Public health indicators in a wealthy city

In view of the seriously deficient and uneven access to health provision in US society, the present section evaluates the extent and characteristics of ill health in Houston. In particular, cancer risks and cancer death from respiratory causes were looked into in detail because of both notoriously high rates and possible links of the disease to environmental conditions in the city. Unless stated otherwise, the documentary information used has been drawn from the annual reports by the City of Houston Health and Human Services Department. This quantitative assessment was very useful for determining the degree of the problem, comparing it to that in other cities, and above all for signalling the contrast between increasing wealth but poor public health indicators. As with any official health statistics, and despite their undeniable value, because they describe the overall health trends in Houston, these are limited in that they are classified only by demographic characteristics, that is, age, sex and ethnicity, and do not include socioeconomic and spatial variables of the deceased or diseased.

General mortality in Houston for each year examined since 1984 was higher than the corresponding US mortality rates.[3] Whilst US age-adjusted death rates have decreased regularly, in Houston mortality trends have remained consistently high (see Table 6.2). The mortality rate for the Houston inner city is a striking aspect of public health in a world city. While infant mortality rates in the

Table 6.2 Age-adjusted overall death rates,* Houston and the US, 1984–90

Year	*Houston*	*US*
1984	626.3	545.9
1985	609.3	546.1
1986	597.8	542.7
1987	598.0	535.5
1988	622.7	536.3
1989	612.0	523.0
1990	695.1	N/A

*Deaths per 100 000 population.

inner city were comparable to those in Third World countries, ten citizens of Houston appeared on the annual list of the 400 richest Americans (Boisseau, 1990).

The infant mortality rate, which includes infants from birth to one year of age, is one of the best indicators of quality of life because babies in this group are particularly sensitive to social, economic, and medical care changes (Boone, 1989). Between 1978 and 1985, Houston, the fifth largest US city, had the highest infant mortality rate (for example, in 1985 Washington, DC had the highest infant mortality rate in the US, at 20.8 per 1000 live births; Detroit, 19.9; Philadelphia, 17.4; New York, 12.8; Houston, 11.2; Dallas, 11.1; Phoenix, 9.9; San Diego, 9.3) (US National Center for Health Statistics, 1987, cited in Boone, 1989). Between 1988 and 1990, infant mortality rates for Houston were also significantly higher than those for Texas and the US (see Table 6.3). In 1988, Houston infant deaths accounted for 1.1 per cent of all infant deaths in the entire US.

Table 6.3 Infant mortality rates* in Houston, Texas and the US, 1988–90

Year	*Houston*	*Texas*	*USA*
1988	11.4	9.1	9.9
1989	11.1	8.0	10.0
1990	9.3	N/A	N/A

*Deaths per 100 000 population.

Infant mortality is even higher in Houston's inner city, a problem characteristic of many industrial and metropolitan cities. In areas of Chicago, Detroit, New York City and Houston, the infant mortality rates approach those found in Third World countries (Sorelle, 1990d). In Houston's inner city, the infant mortality rate in 1989 was a stunning 20.0 per 1000 live births. This was worse than the rates in 25 other nations, including Bulgaria (*Houston Post*, 1990a). Based on their high number of infant deaths, Jamaica (18 for every 1000 live births) and Chile (19) received emergency funds from UNICEF in 1989. In light of this, if, for example, the Fifth Ward neighbourhood in Houston inner-city downtown were a country of the Third World, it would, paradoxically, qualify for aid from UNICEF (*Houston Post*, 1990b).

According to public health records, infant mortality rates vary greatly according to ethnic group. For both infant mortality and life expectancy the statistics are far more favourable for Whites than for Blacks. In 1990, as in previous years, Blacks had the highest infant mortality rate in Houston at 16.9 per cent; the rate for Whites was only 5.8 per cent, for Hispanics 6.7 per cent and others 6.7 per cent (Rich, 1990).[4] While Whites born in 1989 had a life expectancy of 75.9, the life expectancy of Blacks was just 69.7 years.

Moreover, childhood mortality is very high in Houston: in 1989 there were 35.1 deaths per 100 000 population aged 1–14 years; and in 1990 there were 49.5. The 1988 rate for maternal deaths (8.2 per 100 000 live births) was slightly higher than the US rate (7 per 100 000 live births). Although no single cause for the deaths has been pinpointed, maternal mortality has been said to correspond to Houston's high infant mortality rate and that the maternal mortality rates for both groups are directly related to deficient pre-natal care (Sorelle, 1990c). Public hospitals in Houston usually have higher rates of maternal mortality than private hospitals.

General mortality in Houston

The five leading causes of death in Houston listed in descending order during 1990 were heart disease,[5] malignant neoplasm (cancer),[6] cerebrovascular diseases, accidents and adverse effects, and homicide and legal intervention.[7] Together they accounted for 66.1 per cent of all Houston deaths. Comparative data on mortality disease have shown evidentially that ill health may well be a problem in Houston. Indeed, heart disease rates for Houston for 1986–90 were consistently higher than those of the US age-adjusted rates for the same period (see Figure 6.1). Also the 1986–90 age-adjusted death rates for cerebrovascular disease in Houston were much higher than the corresponding US rates, although both showed a modest but steady decline during the four-year period (see Figure 6.2).

The incidence of communicable notifiable diseases (56 in total), such as tuberculosis, measles and sexually transmitted diseases, is much higher in Houston than the national average and the rate of death due to such diseases is three times the national average (Children at Risk Committee, 1990). According to the Center for Disease Control (CDC), Houston represented the fifth largest AIDS case-load of all metropolitan areas nationwide after New York City, Los Angeles, San Francisco and Miami (City of Houston Health and Human Services Department, 1991). Most infections other than AIDS are actually preventable or easily curable with present

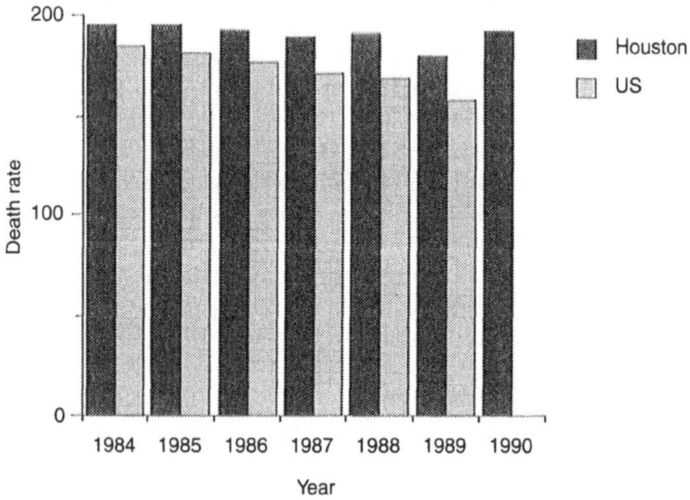

Figure 6.1 Age-adjusted heart disease death rates,* Houston and the US, 1984–90**

*Deaths per 100 000 population; ** death rate not available for the US in 1990.

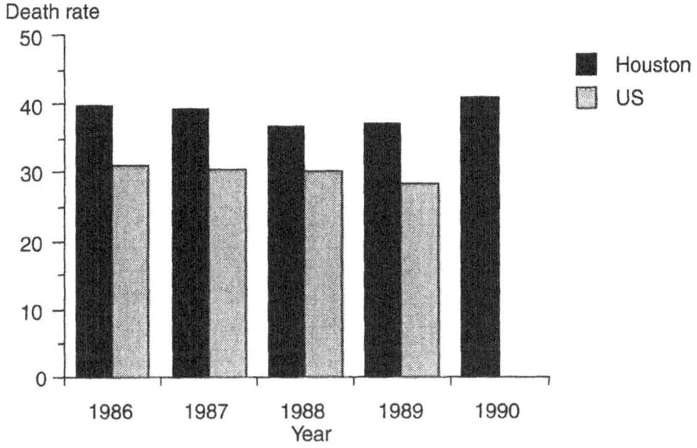

Figure 6.2 Age-adjusted cerebrovascular disease death rates,* Houston and US, 1986–90**
*Deaths per 100 000 population; **death rate not available for the US in 1990.

treatment. However, the wait for appointments in the city's tuberculosis clinics during the 1989–90 school year was up to 90 days. Apart from lack of availability of and accessibility to services, which prevent appropriate care from reaching those who need it, there is also a lack of Spanish-speaking nurses in the city's tuberculosis clinics. This is the case in spite of the fact that 50 per cent of those presenting for treatment are patients of Hispanic origin.

The incidence of tuberculosis in Houston has traditionally been remarkably higher than national rates. During the 1990 period, tuberculosis rates in Houston doubled to 20.4 compared to the corresponding US 10.4 incidence per 100 000 population. Children under 15 years of age represented 6.4 per cent and 7.6 per cent of the tuberculosis cases reported during 1989 and 1990 respectively (City of Houston Health and Human Services Department, 1989a, 1991).[8]

Measles is a highly communicable infectious disease caused by the measles virus. Since the vaccine was introduced in the US in 1963, the reported incidence of measles has decreased by 99 per cent. None the less, the number of cases in the US began to rise exponentially in 1989. In 1990 there were almost 28 000 cases and 97 deaths. Houston, for example, experienced an extensive measles outbreak from early October 1988 to late September 1989. The virus has persisted in unimmunized children living in sizeable pockets of urban poverty (Fenner and White, 1976). Measles is a preventable disease but many community health centres could not obtain sufficient supplies to meet the needs of all their patients because of high prices per dose vaccine (Knight, 1991). And of the overall outbreak in Houston, 30 per cent or 526 cases were preventable. Moreover, there is a clear ethnic gradient with Blacks and Hispanics showing the highest rates (in 1990 the rates were 45.5 per cent and 43.2 per cent respectively; 6.8 per cent White, and 4.5 per cent Other).

Respiratory cancer mortality and risk

Malignant neoplasm, commonly known as cancer, is the second leading cause of death in Houston. From 1984 to 1989 the Houston cancer age-adjusted mortality rates were significantly higher than the corresponding US rates; the 1990 rate is remarkably high (see Figure 6.3). The childhood cancer mortality rate had been stable between 1984 and 1988 at 3.0 per 100 000 population. However, those for 1989 and 1990 are very high (4.0 and 5.4 per 100 000 population respectively).

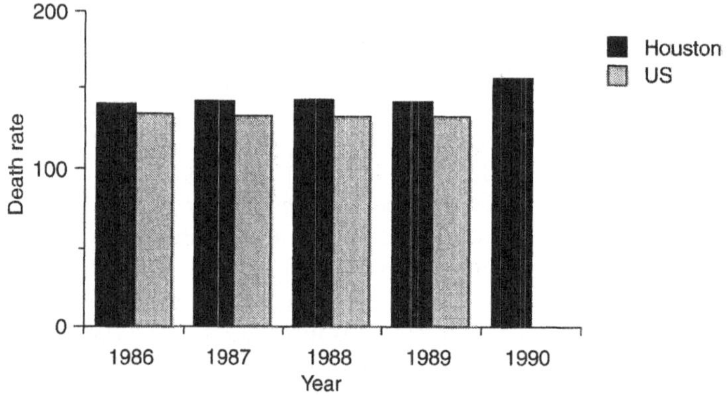

Figure 6.3 Age-adjusted cancer death rates,* Houston and the US, 1986–90**
*Incidence per 100 000 population; **death rate not available for the US in 1990.

Respiratory cancer is the leading sub-category of the disease to cause death in Houston (it represents more than 30 per cent of all cancer deaths; it is followed by mortality from cancers of the digestive organs and peritoneum, of all other and unspecified sites, of the genital organs, of the breast, and of the lymphatic and hematopoietic tissues other than leukaemia cancers). Respiratory cancer death rates in Houston are remarkably higher than the US national rates (see Table 6.4). Compared to other causes of cancer-associated deaths, the prevalence of respiratory cancer is highest among men, 37 per cent, although it is also very high among women, 23 per cent, second only to digestive cancer with 24 per cent.

General cancer death rates have been attributed mainly to ethnicity, but differences in respiratory death cancer rates in particular have been attributed to gender. In fact, neither explanation on its own seems plausible in face of its narrowness. Each ignores a wide range of social and other conditions of the deceased, including exposure to environmental pollution.

MacDonald (1976), whose early study is apparently the only one to have linked air pollution and mortality in Houston, substantially addressed the role of environmental industrial pollution in cancer mortality in Houston. She analysed all death certificates in Houston from all causes since 1940 to 1975 and re-coded them according to the International Classification of Diseases 1955 rubrics.[9]

Table 6.4 Age-adjusted respiratory cancer death rates,* the US and Houston, 1984–90

Year of death	*Houston*	*US*
1984	44.9	38.4
1985	40.9	38.8
1986	40.5	38.3
1987	44.3	39.3
1988	44.9	40.6
1989	45.5	40.3
1990	49.8	N/A

* Incidence per 100 000 population.

Source: adapted from City of Houston Health and Human Services Department I 1989a, p. 5-3; and 1991, p. 140.

A significant relationship was found between industrial ambient pollution and the pattern of cancer mortality, accounting for demographic factors, addresses and exposure to atmospheric pollutants in 15 regions within the city grouped around air pollution sample collection stations. These results are very important because cancer mortality was convincingly correlated with local variations in air pollution.

The addition of relevant additional indicators in present recording of public health, such as the address of the deceased in relation to sources of pollution, could definitely help to ascertain more accurately the causes of death in contemporary Houston. This seems the most appropriate method, taking into consideration that air pollution in Houston may easily reach dangerous levels and that the usual level of pollutants has been moderate to high (Chapter 5). Hence, it might well be that the high levels of air pollution exacerbated high respiratory cancer death rates, which have prevailed as part of this global city.

Estimates of cancer risks from outdoor exposure to airborne pollutants have been expressed as cancer risk in excess lifetime, individual cancer risks and nationwide annual cancer cases (US EPA, 1990b). The maximum lifetime individual risks estimate is 1×10^{-4} (1 chance in 10 000 of contracting cancer). Maximum lifetime individual risk levels exceeding 1×10^{-4} were reported for multi-pollutant exposures from such sources as those found in close proximity to major chemical factories, waste-oil incinerators, hazardous waste incinerators, municipal landfill sites, and publicly owned treatment works (all of which abound in Houston. Based on 90 pollutants and over 60 source categories examined, the nationwide annual cancer incidence was estimated to be between 1700 and 2700 cases per year. This is equivalent to between 7.2 and 11.3 cancer cases per year per million population. Using a total 1986 US population of 240 million, it was estimated that approximately 500 to 900 more cancer cases will occur per year (US EPA, 1990b, pp. 4-1, 4-2). Further, the risk of skin cancer has risen dramatically. The damage to stratospheric ozone by exhaust and other ground toxic emissions has been pinpointed as a main source (Kripke, 1988, 1989; Sobel, 1979).

There are thousands of airborne chemicals that are potentially carcinogenic, but for which there is neither limit exposure nor health effects data (more than 2800 compounds have been identified as existing in the atmosphere;). For example, the highly carcinogenic properties of some air pollutants are well known. It is clear that particles derived from diesel exhaust have a greater effect on biological processes, including lung-tumour development, than those from unleaded fuel exhaust (Department of Health, 1995b). Studies in the UK have revealed that potential damage has been seen following exposure of animals to some combinations of pollutants (ibid.). Of the 90 pollutants evaluated in the EPA cancer risks study, 12 account for over 90 per cent of total annual cancer incidence (US EPA, 1990b, p. ES-3). However, reliable quantitative emission estimates remain unavailable for many potentially important source categories. The lack of data for these pollutants and source categories could result in a significant underestimate of cancer risk.

Both mobile and stationary sources of emission have been found to contribute significantly to the total US nationwide annual incidence of cancer. Considering both direct emissions to the atmosphere and secondary formation, mobile sources had been estimated to be responsible for approximately 58 per cent and stationary sources approximately 42 per cent of total annual cancer incidence in the US or worldwide. The relative contributions of point and area sources to total area-wide lifetime individual risks were consistent with the character of a study of six US geographic locales (US EPA, 1990b). In five less industrialized cities, the chemical 1,3-butadiene, a recognized major carcinogenic, was estimated to contribute between 6 per cent and 24 per cent of the total cancer incidence, all attributable to motor vehicles. However, in the sixth, a heavily industrialized city, over 48 per cent of the total cancer incidence were attributed to 1,3-butadiene. Of the 1,3-butadiene-related cancer incidence in this city, over 80 per cent was attributed to chemical manufacturing plants and less than 20 per cent to motor vehicles (US EPA, 1990b pp. ES-1 and 4–10).

Bearing in mind that due to the effect of airborne pollutants, estimated excess lifetime individual cancer risks and estimated cancer incidence are very high in the US; that in Houston, actual mortality cancer rates are higher than the national average; and that levels of air pollution in the city may reach high levels, it is reasonable to hypothesize that there is a link between the state of the environment and the state of health of the population in Houston.

Epidemiological studies of the type carried out by MacDonald (1976) have been extremely helpful in initiating the study of the relationship between air pollution and cancer mortality in industrialized cities.

Access to the health care system

A world reputation for its 'medical industry' (Greater Houston Partnership, 1995/96), for the availability of the best health services but with access to most of them denied to the less wealthy due to their high price, and a deplorable comparative public health rating are all factors that coexist in this global city. Bearing in mind both that a primary objective for surveying the local residents was to reveal the factors which caused child ill health in a world city and that provision of health care is insufficient in this wealthy place, we examined how the type of medical care and the average expenses of the household affect the state of child health.

The APCH examined whether households used private or public health care and how this fact affected their children's health. The survey has shown that more private (49 per cent) than public (36 per cent) medical provision was used by the surveyed households (n = 300), but households also used both sectors (15 per cent). However, while high-income households spent significantly more financial resources on monthly private health care ($350) than low-income households ($50), the difference in the extent of child ill health between the two types of household was not as large as one might have expected (21 per cent and 28 per cent respectively). The incidence of child ill health was, therefore, high in households which approached public (74 per cent; n = 108), private (63 per cent; n = 146), and also both types (67 per cent; n = 46) of services to obtain health care.

Yet, analysis of access to health care provision alone does not seem to be enough. Therefore, the amount the household spent on health care was also assessed to clarify the role of the health care structures in the incidence of child ill health. The household survey in the four clusters found that, not surprisingly in a market-oriented medical system, many low-income households (29 per cent) reported that it was 'extremely difficult to pay for health care'

(11 per cent in high-income households). The sums of money that households allocated to medical expenses varied greatly and, in fact, among low-income households, monthly expenditure on health amounted to about 5 per cent of the average monthly income ($1250), but, in higher-income households, about 11 per cent of their monthly income ($3300) was spent on health care needs (see Table 6.1). Even so, no significant statistical correlation was found between health spending and the extent of child ill health in low- and high-income households that would sufficiently justify using this variable to explain the large incidence of child ill health reported by the residents.

As to geographic analysis, health provision in the two locations – close to and far from the sources of chemical emission – was similarly distributed with public health facilities being accessed by 33 per cent in the Ship Channel area and 39 per cent in the Southwest; private provision was used by 47 per cent in the Ship Channel area and 50 percent in the Southwest. In the Ship Channel are, 20 per cent of households used either private or public medicine; 11 per cent did so in the Southwest (fewer households used public provisions in the Southwest because there was no public city clinic in this part of the city). However, significantly, only in the less polluted Southwest area did the amount that the household spent on health care make a difference to the state of child health ($P < 0.04$). Evidently, the amount allocated to health cannot, on its own, predict the whole variation of child ill health found in the sample.

Child ill health was reported in 80 per cent of households that used public health facilities; in 69 per cent that used both public and private health care; and in 49 per cent that used private medical provision only (it is noteworthy that the last is also very high, a point which will be scrutinized in Chapter 7). These variables are strongly associated in the less polluted area ($P > 0.001$). In the more polluted area, on the other hand, the incidence of child ill health was distributed in quite a different pattern and there is no sign of a statistically significant association between the extent of child ill health and the type of health provision. First, households that accessed only public health facilities, or used public and private medical care, reported the same incidence of ill health (67 per cent). Second, a high 79 per cent of the households which only used private medicine reported child ill health (see Figure 6.4).

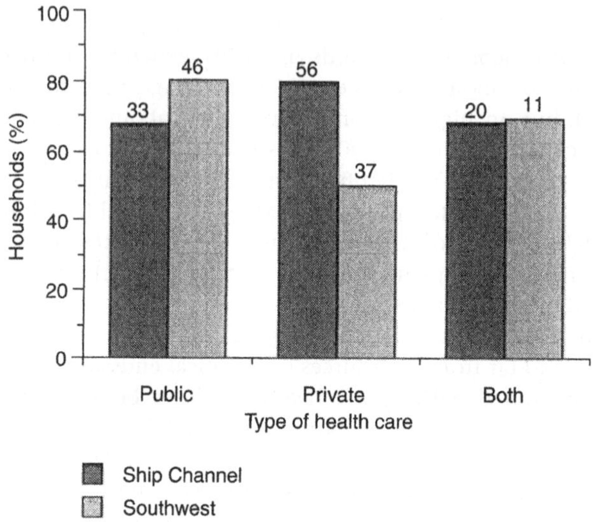

Figure 6.4 Distribution of reported child ill health by type of health care

In summary, the type of health care provision and the household spending on health care may account for the incidence of child ill health only in the less polluted area. The findings so far point to the clear inadequacies of the US profit-oriented health and welfare system and, also, to the possible presence of additional factors which affect the state of child health in Houston.

Occupational class and health variation

Chapter 2 established that social class is a composite generalization that on its own cannot explain health outcomes. Therefore, in order not to confuse the issue of causality, the focus of the analysis of social class and ill health has been on the type of occupation in which the mother was engaged. Importantly, the book examines the mother's, and not the father's, occupation as an indicator of socioeconomic status of the household, access to health services, dampness and demographic factors. According to the City of Houston

Health and Human Services Department annual report (II, 1989b, pp. xv–xvii), the percentage of households headed by single mothers in Houston was approximately 13 per cent, which is remarkably high. Indeed, had the book focused on the occupation of the father, as most studies do, there would have been problems because there was no father figure in 21 per cent of the surveyed households. Moreover, mothers were the focus of reporting because, generally, they spend more time with their children and often they are more aware than fathers of their children's health needs. Also, mothers usually give more explicit and concise information about child health and about the social circumstances of the household. Finally, by focusing on the mother's, and not the father's, occupation, I want to convey that the activity of the mother is economically and socially of great relevance to the functioning of the household.[10] The APCHS in Houston found that 50 per cent of household income variation was associated with the occupation of the women ($P = 0.000$); this supports the choice of the mother's occupation as a class indicator for analytical purposes.

In the enquiry, the mother's occupational class has been classified under five categories. The categorization only broadly corresponds with the social class categories of the British Registrar-General (BRG, 1971 in Black et al., 1982; see Chapter 2 for the I to V categories). The main reason which prevented me from adhering strictly to the same categories of the BRG was that the traditional classification in the BRG only allows for employed individuals who are defined as economically active workers. As a result, mothers who are not formally employed, but have, for example, full-time unpaid work at home, are automatically excluded from the BRG classification. The premise was thus that housewives in the surveyed households played active social and economic roles. Therefore, the 'full-time domestic occupation' category was added to those that were chosen from the BRG. Housework requires manual as well as non-manual skills and therefore is difficult to categorize. Despite the fact that there are also unemployed mothers among high-income households, the occupational domestic category has the lowest status on the scale (I have my own reservations on this position). Initially, occupational classes were classified in six categories. However, because only a few mothers were 'voluntary and students', this category was collapsed into 'professional'. The final occupational classification used in the survey is thought to be most appropriate to analyse the population studied:

I. Professional. Highly skilled and with the highest social status. Included are occupations that require academic qualifications such as medical, technical, scientific, artistic, student, and voluntary work.
II. Intermediate. For inclusion in this group, skills are needed but it is lower in social status. For example, it embraces managerial positions, nursing, school teachers.
III. Semi-skilled. Manual and non-manual. Mainly found in the service sector, such as shop assistants, clerical workers.
IV. Unskilled. Paid domestic work, for example, gardener, and factory machine operator.
V. Domestic. Mothers who are not in paid employment. Their main occupation may consist of unpaid full-time housework, child care, and more.

The incidence of ill health, which corresponded to each occupational class, disregarding any further qualifications, is displayed in Table 6.5. In households where the mother worked in a professional capacity, child ill health was, as expected, lower than in the rest of the surveyed households. The higher incidence of child ill health was found in households of lower occupational status, and this fully agrees with the influential argument that social inequality is the main trigger of ill health (Chapter 2).

Table 6.5 Child health and mother's occupation

Occupation category	*Households in each category**	*Households* ***with*** *reported child ill health, %*	*Households* ***without*** *reported child ill health, %*	*Total, %*
Professional	54	61	39	100
Intermediate	38	74	26	100
Semi-skilled	80	65	35	100
Unskilled	21	67	33	100
Domestic	100	72	28	100
Total	293			

Returning to the issue of housework, these findings, however, should not mislead us into concluding that 'housework', as a skill and occupation, increases the risk of illness. Instead, the state of forced unemployment, which usually accompanies housewives, particularly in low-income households, invariably results in additional poverty, and consequently, in more child ill health. In Houston, 'housewives' in low-income households frequently stay at home because they are immigrants and have a poor knowledge of English; or their illegal residential status in the US impedes them from competing for better-paid jobs. A further obstacle is the lack of affordable child day-care facilities in Houston (Hardikar, personal communication, 1992). In high-income households, conversely, many mothers are not engaged in paid jobs precisely *because* of their high social status and are not expected to work for money. None the less, child ill health in the domestic occupation category was higher only by 11 –14 per cent than for the children of those mothers in the most prestigious occupational categories. Therefore, the distribution of child ill health by occupational class, independent of other conditions in the household, has proven an insufficient indicator to explain the presence of child illness in the total sample.

A negative and robust association was found between the incidence of child ill health in the poor household group and the occupation of the mother (P = 0.000). Among the highest occupational classes, however, there was no clear evidence that high occupational status corresponded with a significant decrease in child ill health. This finding was unexpected because it did not match the claims in the literature. Examination of the particulars of the geographical distribution of the incidence of child ill health threw new light on the findings. While the distribution of occupational classes was similar in the two geographical areas (see Appendix), at the outset, for each occupational category, the incidence of child ill health was higher in households in the Ship Channel area, near the sources of industrial pollution, than in the Southwest (see Figure 6.5).

Further, with respect only to low-income households, but both for those in the Ship Channel area and in the Southwest, a similar incidence of child ill health was reported (for intermediate, semi-skilled, unskilled and domestic occupations, 83 per cent, 64 per cent, 71 per cent and 77 per cent respectively in the Ship Channel area; 100 per cent, 72 per cent, 67 per cent, and 80 per cent respectively

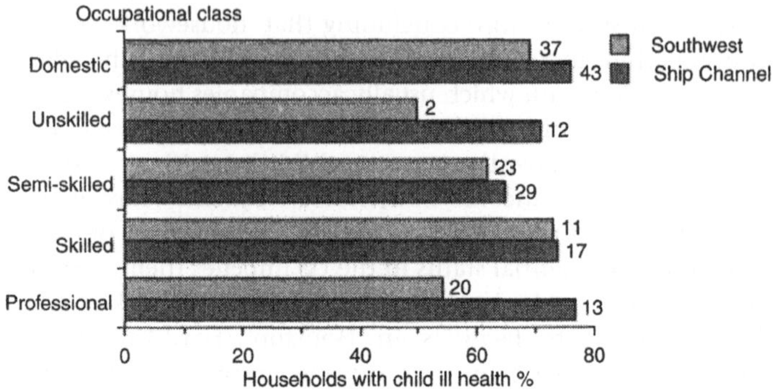

Figure 6.5 Spatial distribution of child ill health accounting for mother's occupation

in the Southwest). Similarity in the geographical distribution of these households with incidence of ill health could be attributed, primarily, to the adverse effects of poverty. However, the geographic variation of the incidence of reported child ill health was particularly pronounced among highest-income households (see Figure 6.6). In the professional group, 23 per cent more children were suffering ill health in the Ship Channel than in the Southwest. In the domestic category, 35 per cent more households in the Ship Channel than in the Southwest reported ill children. The percentage of child ill health reported by mothers in semi-skilled professions was 46 per cent higher in the Ship Channel. Very few mothers in high-income households in the Ship Channel worked in unskilled occupations (n = 3) (one worked in the Southwest). Reported ill health was 67 per cent in the Ship Channel and 0 per cent in the Southwest. There was thus a strong indication that other factors, which varied by household geographical location, were relevant in the association.

In summary, the effect that occupational class had on the variation of child ill health corresponded with the general income group. None the less, this combined effect was consistently modified by the spatial distribution of the household. Since location *per se* cannot cause ill health (see Chapter 2), the presence of geographical factors which may have varied by location of the household seemed the

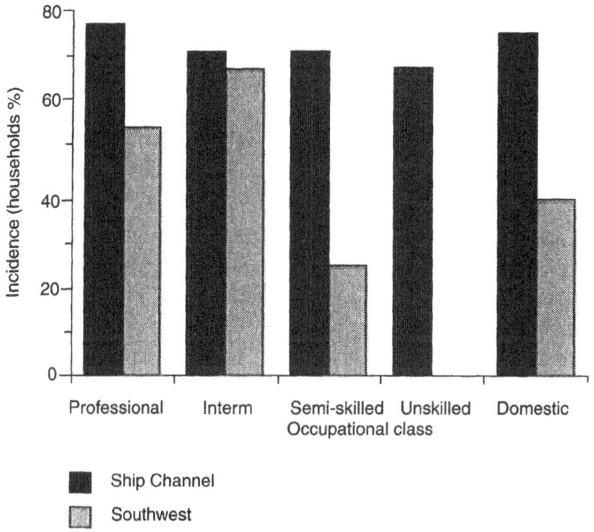

Figure 6.6 Spatial distribution of child ill health in high-income households accounting only for mother's occupation

most likely explanation. Before finally moving to examine the effect of spatial variation, the role of household demographic characteristics in the incidence of child ill health will be assessed.

The role of demography and family structure

Two structural family factors, the number of children in the household and parental unit composition, were assessed for their prospective effects on the state of child health. Household size is apparently highly interrelated with ill health in that infections are more likely to be spread within to larger households, with the risk of child illness increasing as the number of children in the household rises (Butler and Golding, 1986; Leeder et. al., 1976; Harlap et al., 1973). However, the Houston APCHS shows that the number of children who live in the household, and whether one or two parents headed the household, correlates, above all, with household income, and ill health is most likely to be attributable to this factor.

The chance of finding ill children increased remarkably as the number of children in the household rose from one to two. Overall, there was 20 per cent more child ill health in two- (73 per cent) than in one-child households (53 per cent; n = 75). In two-, three-, and four- and above child households (n = 105, n = 72, and n = 48 respectively), the incidence of ill health was very high but it did not change proportionate to the rising number of children (73 per cent, 71 per cent, and 75 per cent respectively) (see Figure 6.7).

Undoubtedly, the fact of the definite jump in incidence of illness between single-child households when compared with households with two or more children was significant. The data showed that the extent of ill health was considerably higher in low- as opposed to high-income households, but that rising number of children did not seem substantially and directly to affect incidence (see Figure 6.8).

The combined effect of the number of resident children and geographical location of the household exposure to atmospheric pollutants was assessed by comparing the relationship in the air-polluted Ship Channel and in the less polluted Southwest areas. Overall, child ill health was very high in the Ship Channel area particularly

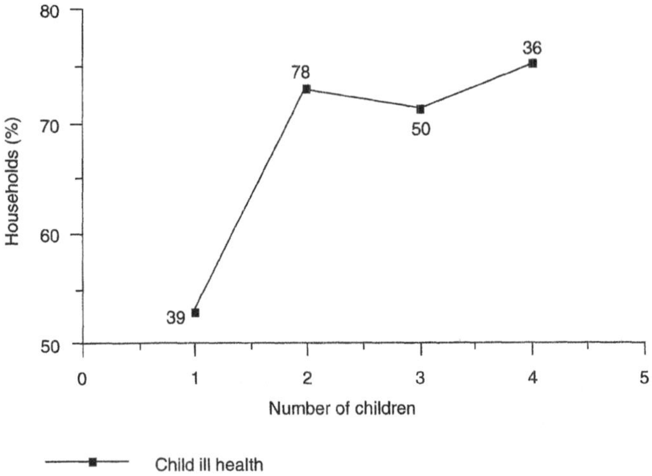

Figure 6.7 Incidence of child ill health by number of siblings in the household

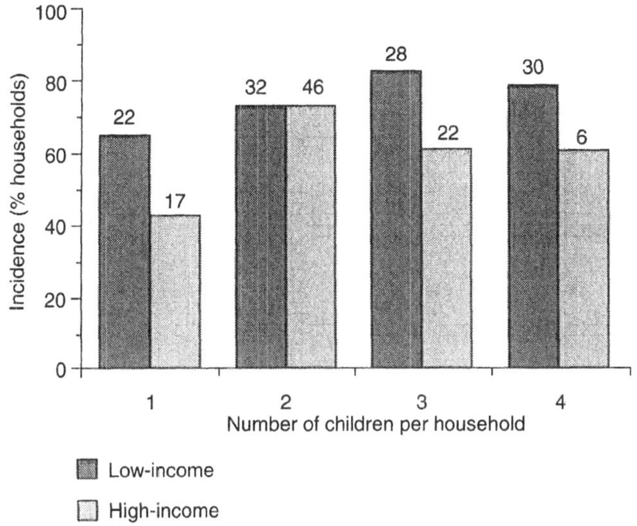

Figure 6.8 Distribution of child ill health in low- and high-income households by number of siblings**
(1 missing case); **category 4 on the horizontal axis means 4 or more children.

for one- and three-child households; it was however slightly higher in the Southwest for two- and four- and above child households (as shown before; see Figure 6.9). Higher incidence in the less polluted area might be attributed to poverty. Significantly, only in the Southwest are the variables child ill health, the household group (P = 0.009), and the number of children in the household (P = 0.000) statistically associated.

Focusing on the incidence of child ill health in the highest-income households only was illuminating. It was found that child ill health was 40 per cent, 10 per cent, and 38 per cent higher in one-, two-, and three-child households respectively in the Ship Channel than in the Southwest (see Figure 6.10).

The number of children in the household and the socioeconomic status of the household thus acted as partial indicators of the extent of child illness. Yet the extent of the incidence of child ill health in the Ship Channel area still remained unexplained by these indicators

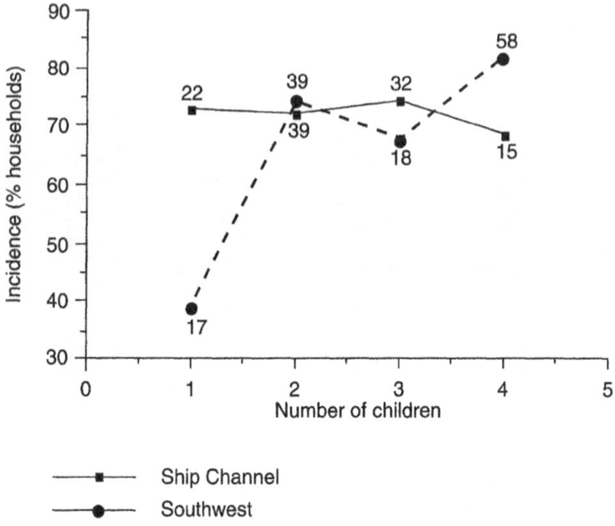

Figure 6.9 Distribution of child ill health by spatial location of the household and number of siblings

alone. These findings signalled that there were causal factors, which varied by location and were influential in the pattern found.

As previously stated, single- (female-)headed households in the sample population were very numerous (21 per cent) and the survey found that there was no correlation between parental unit structure *per se* and the state of child ill health, which was only 4 per cent higher in single- than in two-parent households (see Figure 6.11). In general, this finding agrees with Blackman et al. (1989). In a housing and health study in West Belfast, they found no significant difference in the health of the children of one-parent compared to two-parent households.

As for average household income, half of all lone-parent households (49 per cent) reported low income whereas only a small fraction (8 per cent) earned high incomes (20 per cent reported incomes in the second low-income bracket, and 22 per cent in the third) (see Figure 6.12).

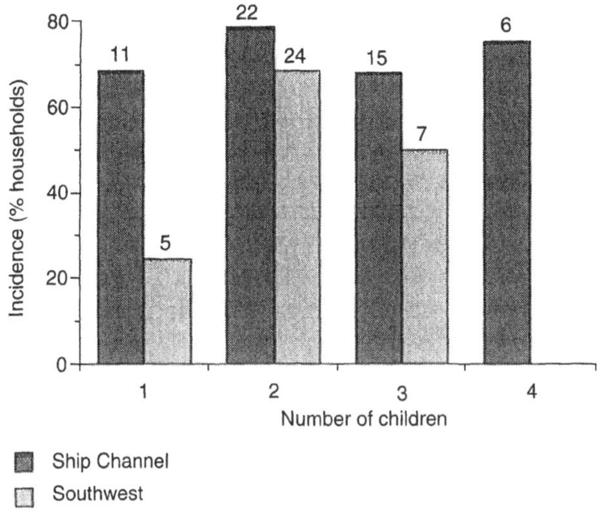

Figure 6.10 Spatial distribution of ill health of children in high-income households accounting for number of siblings

These figures closely agree with the Children at Risk Committee's (1990) report on poverty and single parenthood in the state of Texas, which found that 42 per cent of single-parent households lived below the poverty line and an estimated 76.8 per cent of all single mothers under the age of 25 lived in poverty.

When parental unit structure was examined together with income brackets and spatial location of the household, the effects of parental structure became more obvious in two ways. First, clearly, there was more child ill health in low- than in high-income single-parent households (22 per cent). However, incidence of child ill health was not significantly lower in low-income two-parent households (see Figure 6.13). In the rest of the households, child ill health was generally lower in one- than in two-parent households. There was clearly a causal relation between the economic resources of the household and parental structure, and this particular relation seemed to have an effect on the state of health of the children.

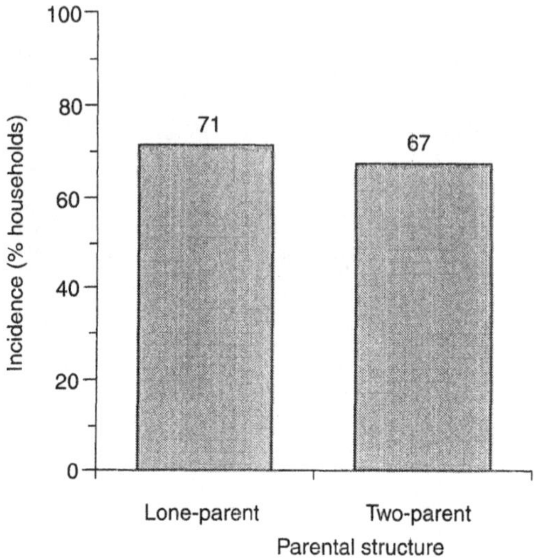

Figure 6.11 Incidence of child ill health by household parental structure (lone-parent households n = 63; two-parent households n = 237)

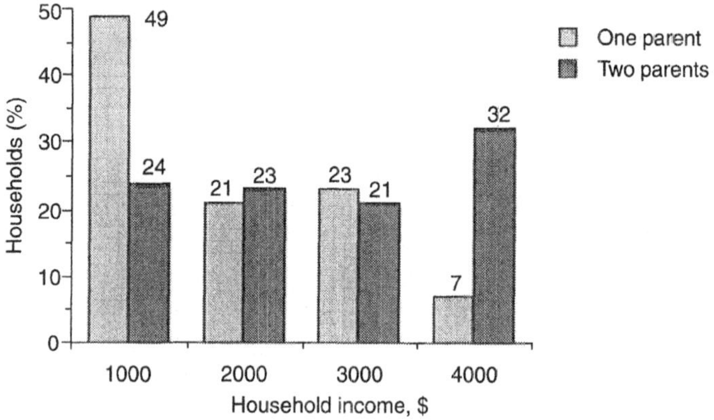

Figure 6.12 Household level of income and parental structure

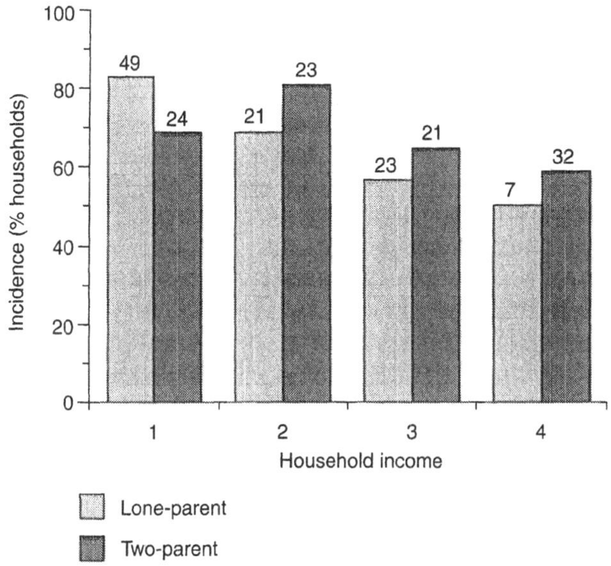

Figure 6.13 Variation of reported child ill health by parental structure and income levels*
*As per figure 6.12, US$1000, US$2000, US$3000, US$4000 average monthly income.

Second, the distribution of child ill health in single- and two-parent low-income households was very similar in the Ship Channel and the Southwest residential areas (for lone-parent, 70 per cent and 78 per cent; for two-parent, 78 per cent and 78 per cent). However, in high-income households, more child ill health was found in the Ship Channel than in the Southwest residential areas in both single- and two-parent households (see Figure 6.14). These findings clearly indicated the strong influence of local variables.

Dampness and ill health

It is important to examine the effect of dampness on child health because dampness is one of the most common health hazards associated with poor housing conditions and respiratory ailments. Dampness seems seriously to affect the health of children, particularly

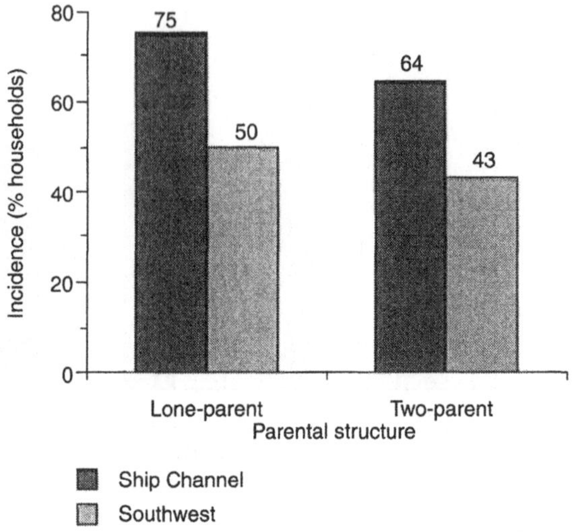

Figure 6.14 Distribution of child ill health in high-income households (n = 150) accounting for parental structure

in relation to the level of respiratory/bronchial symptoms. Headaches, diarrheoa, aches and pains have all been more commonly reported among children living in damp rather than dry dwellings (Yang et al., 1997; Hart et al., 1986). Mould growth was pinpointed as the factor responsible for the significantly worse state of health of children in damp houses; spores germinating under moist conditions may enter the respiratory tract, causing bronchial and asthmatic symptoms including fever, tiredness and lethargy. Children then have less resistance to the allergens and are more vulnerable. Also, allergic reactions may occur to the house dust mites and storage mites that multiply in damp conditions. The mycotoxins, or mould given off by fungi, may get into the mouth or nose and be swallowed, causing stomach upsets as well (Hart, 1986).

This aspect of housing was explored in particular because of both very hot and humid climatic conditions in Houston and the widespread use of artificial air conditioning in most buildings (see

Chapter 4). Here, a note on central air conditioning is necessary. There are reasons for the higher percentage of households with central air conditioning in the Southwest than in the Ship Channel area (see Chapter 4). In the Southwest, large expensive houses have been built more recently and therefore have central air conditioning. Most of those on low incomes in the Southwest live in rented apartments which always have central air conditioning and it can be used at no extra charge. In the Ship Channel area, on the other hand, many low-income families live in cheap and generally old small houses that do not have this facility. Moreover, many of the most expensive houses in the Ship Channel area are traditionally and elegantly old and do not have central air conditioning; the many new homes are fully air conditioned. Air conditioners such as electric units were usually found in these homes. Importantly, lack of natural air inside air-conditioned buildings can be aggravated by the fact that buildings are kept closed most of the time to maintain a pleasant temperature and to prevent entry of mosquitoes. This results in inadequate room ventilation, which may in turn produce residual humidity. For dwellings without central air conditioning the situation is not much better.

Altogether, 19 per cent of the sample (n = 300) reported dampness and the distribution was very similar in the polluted Ship Channel

Table 6.6 Spatial distribution of child health accounting for income and dampness in the house

Households	*Total households, no.**	*Ship Channel, no.*	*Southwest, no.*	*Child ill health, %*	*No child ill health, %*
Low income					
Damp in house	34	18	16	82	18
No damp in house	106	57	59	71	29
High income					
Damp in house	25	14	11	67	33
No damp in house	118	61	64	58	42

*15 cases 'did not know' whether there was dampness in the house; 2 observations were missing.

and control areas (see Table 6.6). The incidence of child ill health was particularly high in houses with reported dampness. However, while dampness must certainly have contributed to this incidence, the comparative study revealed that the absence of dampness in the house did not preclude households from reporting a high incidence of health problems (65 per cent of the sample). These findings confirm that dampness is likely to aggravate ill health, particularly in poor households. None the less, in high-income households with no reported dampness, the extent of child ill health was also surprisingly high.

The incidence of child ill health in damp houses (n = 45) was higher in the Ship Channel (70 per cent) than in the Southwest area (60 per cent). Ill health in low-income damp houses in the Ship Channel and in the Southwest was similar and very high (79 per cent and 83 per cent respectively; see Figure 6.15). This could be attributed to social disadvantage. On the other hand, in high-income

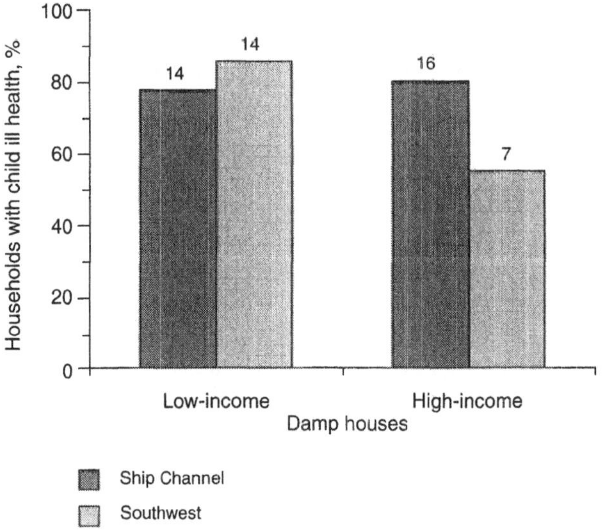

Figure 6.15 Spatial and socioeconomic distribution of child ill health in damp houses

damp houses, the incidence of child ill health was 25 per cent higher in the Ship Channel than in the Southwest area.

There is thus an indication that, in the presence of dampness, the conditions for ill health might have been aggravated by the location of the household.

Conclusion

This chapter has contributed to a deeper understanding of the institutional role of health care in relation to air pollution. A strong correspondence between residents' reported and monitored air pollution has emerged from the enquiry in Chapter 5.

Houston is a city of conspicuous contrasts between wealth and professional excellence on the one hand and industrial hazards and poverty on the other. In addition to outstanding industrial and business development, the medical industry has won international acclaim; a few Houston residents rank among the wealthiest individuals in the US; this global city has all the appearance of a flagship advanced technological centre of the future. However, this chapter has substantively indicated that access to proper medical care is reserved only for those who can pay dearly. Poverty is by no means excluded from Houston. The advantages of economic growth seem to be far outweighed by the reviewed disadvantages. High infant mortality and cancer deaths give ample cause for concern, as does the remarkably high number of days annually classified as injurious to health according to the US pollutant safety standards. Notwithstanding that Houston stands out as an economically prosperous US city, with excellent medical facilities and specialized health services, the chapter has demonstrated that provision of health care for all is severely constrained by major structural obstacles which are rooted in the market orientation of the medical sector. Analysis of public health indicators has shown relatively poor achievement in Houston, which is even more surprising when one bears in mind that US health achievements lag considerably behind those of the other six major world economies. There is also an indication that significantly high levels of mortality and, particularly, of respiratory cancer in the Houston area were related to the adverse effects of exposure to industrial emissions.

The distribution of reported child ill health and the presence of air pollution in households in the less polluted area always correlates, as expected, with household socioeconomic and demographic circumstances. Indeed, in the Southwest, the extent of child ill health diminished if the household had higher income, fewer children, higher occupational status, and two parents. In contrast, child-health variation in the Ship Channel area did not change consistently with different social and demographic household circumstances such as household income, number of resident children, mother's occupation, dampness and parental structure. In summary, poverty always has an active role in triggering ill health; however, it was the large incidence of reported child ill health in high- compared to low-income households that hinted at additional risk factors of a spatial character ($P = 0.014$). This was significantly expressed in the geographic distribution of child ill health in regard to socioeconomic and demographic variables ($P = 0.001$). The combination of various socioeconomic and demographic circumstances of the household with the geographical location seemed to offer causally relevant information to uncover the origin of the high incidence of child ill health in Houston.

The analysis of the database from the two areas within Houston has exposed the importance of the home's geographical location in relation to industrial sources of air pollution as compared to socioeconomic inequality. The case study has been essential to highlight the presence of localized environmental risk which could not simply be explained by socioeconomic factors. The theoretical analysis has stressed the relevance of structures to determine environmental risk (Chapter 3). Chapter 4 has stressed the interconnections between local development over time with regional and global economies, while Chapter 5 proved the current presence of air pollution with a history. The information from the case study suggests a strong connection between local ecological and residents' health changes. Significantly, residents became the voice against this silent local contamination that has been afflicting their children's health.

Widely held assumptions about the role of socioeconomic structural conditions in the household have proven helpful, but not sufficient, to understand the regularities and irregularities that have emerged from the household survey in Houston. Without denying that poverty is a main cause of ill health, this fundamental assumption should be further qualified in the light of the findings of the book's enquiry. Assessment of the role of additional factors that focus on the individual, such as occupational exposure or tobacco smoking, would be practically irrelevant in the presence of the powerful health effect of outdoor industrial pollution. Such additional externalities cannot alter or invalidate the fact that air pollution is a direct and pervasive trigger of ill health in Houston. Individual rather than collective exposure has been the focus of these approaches, and one cannot but note that the political stance has been one of evading responsibility for ill health.

It is irrefutable that, in the twentieth century, Houston achieved groundbreaking medical developments, a reduction in infant mortality rates, noticeable health status improvements, and longer life expectancy. Economic growth in the developed nations seems in great measure to have facilitated these advances. However, the above discussion is indicative of three additional outcomes. First, only a section of the population is permitted to benefit from the advances in medicine (and this holds true for any city). Second, it is clear that the actual concern for health has lagged well behind that for the desire for economic growth in general. Had health and the environment been protected under the same political commitment as economic growth, there surely would be evidence of significant reduction in the incidence of disease, premature death and cancers. Lastly, a further outcome, and one that concerns us most, is that unregulated economic growth has avoided simultaneous strong environmental protection. The latter has not materialized, and will not, in the short term at least, because the very economic activities that have promoted growth, and financed medical research and development, have represented a main factor for environmental contamination. The next chapter looks into how the regulation of industrial emission is related to these consequences.

Notes

1. Medicare, which began to operate in the 1960s, provides affordable health insurance to elderly Americans. Medicaid was devised to provide health care to the poor. The federal government shares the costs under Medicaid, but the individual states set eligibility criteria and administer the programme.

2. Much like an insurance programme, Medicaid reimburses hospitals, physicians, pharmacies and other health care providers for the medically necessary care they provide to persons covered under the programme.
3. Age-adjusted death rates, as opposed to crude rates, control for the effects of different age structures.
4. Mexican-American infant mortality, however, was found to be unexpectedly low (*Houston Chronicle*, 1990a).
5. The age-adjusted death rate was 188.8 per 100 000 population.
6. For the incidence of cancer in the state of Texas, see the joint publication by the Texas Cancer Council, the Texas Department of Health, the University of Texas M.D. Anderson Cancer Center & Other Institutions (1991).
7. The 1990 rates use the 1990 census. The 1989 and 1990 rates which use population data in the denominator should not be compared.
8. High immigration, particularly from the Mexican border, has contributed to maintain the very high incidence of tuberculosis. Moreover, there is a clear ethnic gradient, with Blacks and Hispanics showing the highest rates (in 1990 the rates were 45.5 per cent and 43.2 per cent respectively; 6.8 per cent White, and 4.5 per cent other).
9. Death certificates were classified by age-adjusted mortality rates from heart disease, cancer and stroke, as well as from all other causes, total respiratory disease and from malignant respiratory disease.
10. In the UK, census, housewives, or full-time unpaid domestic workers, have been grouped under the category 'other economically inactive' despite their economic and social importance. Duncan (1991) has estimated that if the totals of housewives and women of independent means in the UK, both sub-categories of 'other economically inactive' in the census, were added to the economically active total, the economically active census figure for women would almost double.

7
Politics of Protection Betrayal

Introduction

In any modern city, the triggering of an environmental episode takes place because the socially constructed context allows it. Hence, factors other than the purely physical, chemical and biological also intervene in the relationship. Indeed, remarkable levels of industrial air pollution in Houston have necessarily presupposed the development of industrial plants, the use of motor vehicles, and a particularized environmental legislation. Rather than pinpointing industrialization as the ultimate cause of excessive contamination, this chapter argues that lax environmental regulation has accompanied unrelenting economic growth, and that together, these have represented crucial factors in the institutional and practical degradation of the local environment.

The chapter explores the ways in which the industrial emission control legislation has come to an accommodation with, rather than sought to constrain, those same economic trends that since the beginning of the twentieth century have caused unabated environmental degradation and, seemingly, contributed to an increased risk of contracting chronic and deadly diseases. Previous chapters have shown that massive industrialization tightly linked to economic growth and globalization have participated in the creation of concrete forms of urban development and urban degradation. The case study has indicated that, in Houston, children in households located in the proximity of the industrially developed Ship Channel area were more directly affected while others did not suffer equally or were not affected at all. The last chapter showed that low public health achievements in Houston stood in opposition to the internationally renowned medical industry in this world city. The aim of the present chapter is to discuss the extent to which public health has indeed constituted an issue of concern and importance within the environmental legislation that has regulated industrial emissions.

The degree to which the government has fulfilled its role to protect the environment, and hence the health of the city's residents, by means of regulating and controlling emissions may be viewed as an indication of its commitment to improve the basic well-being of the residents. As has been mentioned, the experience of deadly air pollution in US and European cities in the first half of the last century encouraged the passing of limited local legislation or of private litigation. In Los Angeles, smog was a problem from the 1940s, because emissions, principally from motor vehicles, oil refineries and back-yard incinerators, were converted to eye- and nose-irritating pollutants by photochemical reactions. It was not until the California legislation of 1947 that a state law tackled air pollution other than dense smoke, and not until 1952 that the state of Oregon introduced the first comprehensive state air pollution control legislation (Elsom, 1992). In the UK, only after around 4000 people died in the 1952 London sulphur pollution episode was the first Clean Air Act passed in 1956 (Chapter 1). Since then, UK and other European legislation has considerably tightened controls on air pollution from the so-called 'stationary' sources of homes, commerce and industry (Ball and Bell, 1994). Regular monitoring of air pollution in the US and the UK only began in the early 1970s. In Houston, effective nationwide pollution control was repeatedly opposed by the robust industrial lobby. Restraints on rapid growth and caution over the rights of property owners would have been required to deal with oil-related damage to the environment (Pratt, J.A., 1980, p. 243). In the face of increasing air and water contamination from the cumulative impact of half a century of industrial and population growth in Houston (Chapter 4), pollution could no longer simply be ignored and more forceful means of implementing solutions were sought.

The modern non-interventionist environmental federal and state legislation reflects the conflict between the powerful commitment to economic growth, regardless of the consequences, and an attempt to protect the environment from further damage. It has been shown that, historically, regulations to control the emission of toxic materials have been weak and that this is indicative of the prevailing pro-growth forces that have acted incessantly in the region. Policy-makers have traditionally considered strict environmental policy as a threat to the economic success of local firms and global corporations. Local as well as national interests have all shown signs of concern for the international competitiveness of the region. Competitiveness has often been judged on the basis of current levels of trade, investment, or output (OECD, 1997). Hence, policy initiatives that threaten these variables are seen as reducing the ability of the nation-state, or city, to compete on the international stage. This chapter argues that rather than improving environmental sustainability of the city, efforts have been directed either at avoiding more stringent environmental policies that would have imposed unacceptable economic "hardships", or at mitigating pressure that would have grown to reduce the strength and effectiveness of environmental policies. The pressure on the governments of developed countries to implement environmental controls has increased globally to reduce overall levels of carbon dioxide. Yet Houston has evolved as a 'pollution haven', and has stayed like that, despite some more decisive recent attempts by the national government to change this.

The Houston study represented an attempt to address the fairly common tendency in current social and scientific research to present biophysical events as ahistorical and unrelated. We have argued that one basic aspect of persistent environmental degradation is its unique physical and independent character. Recognition of ecological mechanisms is necessary in order to determine the capacity of the environment either to absorb human pollution and to regenerate itself, or to disrupt the atmospheric chemical balance. Yet empirical biophysical phenomena in large urbanized centres have been mediated by society. Chapter 4 connected the deterioration of the local environment over time to past and current growth trends in the oil sector. Chapter 5 has shown that the levels of monitored pollution exceed healthy limits and that residents suffer the consequences. Chapter 6 began to account for the health effects in socioeconomic and health care terms. The present chapter looks into the ability of environmental regulation to affect the pollution and health outcomes. In the multi-faceted model, the institutional analysis is complete when the role of the main regulatory bodies participating in the equation has been discussed. It is assumed that the way that the environmental regulatory bodies work in a city like Houston has definite consequences for the only partial attainment of the national safety standards, and hence for the considerably less than optimal health of the population. The spatial dimension of the problem of urban air pollution is specifically addressed here.

Emission control and Public Health

Preoccupation with health appears to be a major priority of the US Air Quality Act of 1967. It aims 'to protect and enhance the quality of the Nation's air resources so as to promote the public health and welfare and the productive capacity of its population ... to ensure that air pollution problems will, in the future, be controlled in a systematic way' (Bach, 1972, p. 104). Until the 1970s, the Act still left the primary responsibility for air pollution control to the states and local authorities, but required the federal government to intervene if necessary. The protection of health has repeatedly appeared as central to the concerns of the regulation. This point is even more evident in the latest US EPA report, which analyses the ten-year progress between 1987 and 1996 in reductions of emissions (US EPA, 1998). Even though the regulation of air pollution in the state of Texas has not fully conformed to the Clean Air Act (1990), it makes the protection of public health its foremost concern. The dedication to improved health is explicitly articulated in the document:

> The presence in the atmosphere of one or more air contaminants or combinations thereof, in such concentration and of such duration as are or may tend to be injurious to or adversely affect human health or welfare, animal life, ...
>
> (TACB, 1987, p. 2)
>
> ... to ensure that emissions from new and modified sources do not result in adverse effects on public health or welfare.
>
> (ibid., p. 9)

However, the extent to which health protection can actually be achieved depends not only on the concerned words of the legislation, or the technical approaches in use, as the following section shows. Certainly, a number of conditions may preclude the declared commitments to protect health as they appear in the legislation. Among them, for example, are factors such as geographical restrictions in the methods of measuring people's exposure, a corrupt permit procedure, the limitation in the pollutants' modelling methods, flaws in reporting excess emission, and the common inadequacy of the NAAQS to safeguard health. The authorities in Houston or in Texas as a whole have not complied with the targets of the 1990 Clean Air Act. The national air quality standards, particularly for ozone, have been constantly violated. The ozone readings have reached such high levels that residents in Houston have at times been advised to stay indoors to protect their health (Yardley, 2000).

Throughout the 1916–41 period, oil industrialists in Houston had taken some initiatives to control the pollution, mainly of water, resulting from their activity. However, these had little impact on pollution control during the period because, 'while a small group of researchers looked for answers to basic questions involving the treatment of water discharge, a much greater industry-wide research effort developed new techniques of production' (Pratt, J.A., 1980, p. 238). From 1901 until the 1960s, several state agencies – underfunded and understaffed as they were – were at the centre of public initiatives to control air pollution in Houston. Each of the existing agencies reacted to pollution as a minor problem, and 'although pollution sometimes became a threat to public health, local and state public health departments had many more pressing demands on their resources' (ibid., p. 231). Perhaps even more significant than the general lack of resources was the absence in the early years, as throughout most of the century, of even an underfunded public body with a specific and overriding institutional mandate to control pollution.

Pollution control in Houston had been influenced by new federal laws enacted in other cities throughout the 1950s and 1960s. A US national approach to the problem was stimulated by the extreme six-day smog episode in Donora, Pennsylvania in 1948 (which caused 6000 cases of illness and 20 deaths), and also by the later smog and deaths episodes in London in 1952, and in New York in 1953 (Chapter 1). The first national air pollution legislation in the US, the 1955 Air Pollution Act, marked the beginning of limited federal involvement in a policy-initiating role (Elsom, 1992).

In Houston, a gradual emergence of public leadership in pollution matters occurred only in the early 1950s, when a group of citizens who lived near the Houston Ship Channel began a private campaign to reduce pollution through meetings with local industries. Their protests generated pressure for the creation of such an agency and encouraged the formation of an air and water pollution control section in the County Department of Health. This had only limited power to force reduction in discharge. Not until 1961 was the first state pollution control agency created, the State of Texas Water Pollution Control; and in 1965, the Texas Air Control Board (TACB) was brought into force. However, private organizations and the large oil companies retained a strong interest in imposing their own air and water quality standards on these public institutions, and the oil firms did not easily surrender their traditional autonomy in this area. In its early years, 'the board did not hamper the growth of industry or of pollution' (Pratt, J.A., 1980, p. 243). Indeed, programmes of local air pollution control were far from adequate.

In contrast with earlier times, the environmental movement in the US, peaking in 1969 and 1970, provided the political muscle necessary to counter the lobbying by industrialists. It resulted in the enactment of the innovative and sweeping Clean Air Act of 1970, marking the beginning of the present era of pollution control policy (Elsom, 1992). Uniform national air quality standards were set to protect public health and welfare (the NAAQSs; Chapter 5). In 1970 the agency was renamed the Air Pollution Control Office, and began to operate within the newly established Environmental Protection Agency (the US EPA). None the less, quality standards have been flexible. In fact, as a result of non-attainment of required national standards and continued pressure from petroleum industrialists, the Clean Air Act was subsequently relaxed and weakened during the late 1970s (Elsom, 1992). For example, the maximum allowable ozone concentration was increased from 0.08 to 0.012 ppm in 1979 – which is the present NAAQS, and sulphur dioxide emission requirements for major sources were relaxed in 1981 during the Reagan administration.

Permits and technology to control pollution

A main procedure of the emission control regulation to protect public health has consisted of determining whether each proposed source is in compliance with the intent of the Texas Clean Air Act (TACB, 1987). A further rule has been that the permit-authorizing staff will predict the health impact of chemical emissions on the population by using engineering and modelling techniques. They estimate emissions and ambient concentrations employing the applicant's emissions estimates once they have ensured that best available control technology (BACT) has been incorporated into the design of the facility. The cornerstone of the effects evaluation process and the dispersion model is the use of an effects screening level. These measures are chosen to protect against adverse health effects, vegetation effects, materials damage including corrosion, and nuisance conditions such as odour. The short-term (30 minutes) health effects screening level is equal to 1/100th of the most applicable occupational exposure limit. For a long-term (annual average) health effects screening level, 1/1000th of the occupational limit is used. However, if the screening levels are exceeded, this does not necessarily mean that the project cannot be approved. Instead, a more extensive review is initiated (TACB, 1987, p. 5). Furthermore, modelling can prove a very limited method. Although the technique may be sensitive to spatial data, it is not sensitive to the actual health consequences suffered by residents exposed to emissions, or to topographic and climatic factors which may well exacerbate the deleterious effect of toxic emissions. In these circumstances, it is plain that the issue of health safety has not necessarily been the leading priority in the control of industrial emissions. The legislation covering the medical industry operates along similar lines. To what degree, then, is the advance of Houston's medical industry in the interest of protecting public health, and to what degree to promoting its economic growth?

The major type of health problem that the environmental authorities in Houston, and in Texas in general, deal with is odour nuisance, which mostly originates in the vast local chemical industry (Chapter 5). To evaluate these situations, the staff review the literature, rather than rely on other methods of measurement, to find the odour thresholds for the substances in question. If people were exposed to an average concentration of odorous compound equal to the odour threshold concentration, most would notice such an odour. There is, however, an element of professional judgement in assessing whether or not the reported odour nuisance condition will be caused by the emissions from an industrial facility (TACB, 1992d, p. 4), and whether the local residents have been exposed to it. Further, detection of odours may require immediate assessment and sometimes it may be impossible for the authorities to reach the scene fast enough. However, residents, and not professionals, are obviously more capable of assessing this type of odour because they are present at the time of the nuisance and may more easily than non-residents recognize changes in the smell of the air.

How effective, in reality, is the requirement by the system for the polluter to report its own excess emissions to the executive director and to the appropriate local air pollution control agencies? The industrial polluter must do so as soon as possible after any major incident that causes or may cause an excessive emission that contravenes the intent of the Texas Clean Air Act or the regulations of the Board. Notification should identify the cause of the incident and the processes and equipment involved, and should include the date and time of the event, which can be reported up to two weeks after the onset of the upset condition (TACB, 1987). Self-reporting, is, however, flawed because it depends on the goodwill of the owner or operator of a facility. Residents' well-being may easily be adversely affected by this procedure because they may not be informed on time to take preventive action. In fact, hazardous leaks and other chemical accidents have occurred for years in this world city, a situation which, in part, has influenced long-awaited, but only recently achieved, semi-voluntary agreements between the US government and major industrial polluters (Chapter 8; US EPA, 2000).

A further problem of industrial negligence that has become evident in light of environmental regulation is the situation when emissions occurring during major incidents may not be required to comply with the allowable emission levels set by the rules and regulations upon notification. This is the case if, after consultation, a determination is made by the executive director that the emissions were unavoidable and that a shut-down had to be implemented (or other corrective actions taken) as soon as practicable. Also, excessive emissions occurring during start-up or shut-down of processes or during periods of maintenance may be exempt from complying with the allowable emission levels set (US EPA, 2000, p. 16). Clearly, the air may suffer further contamination as a result of such significant regulatory concessions, naturally increasing the risks of adverse health consequences for the city's inhabitants, and particularly for those who live near the sources of toxic emissions.

Finally, sources emitting air contaminants which cannot be controlled or reduced due to lack of technological knowledge may be exempt from the applicable rules and regulations when so determined and ordered by the TACB (TACB, 1987, p. 16); and information of the health effects of a compound often does not require a tightening of controls on permitted facilities that emit it. The reason for this is that it is thought that BACT requirements have usually led to emissions control that is better than merely adequate to meet a screening level. Also, exceptions from a full permit review are granted to some sources where the residual emissions are thought not to have the potential to cause air pollution.

Historically, the pollutants health safety limits imposed by the government have oscillated in response to economic and political pressure from local and global industrialists and entrepreneurs. Traditionally, the relative health risk to a population has been assessed through a combination of the measured concentration of air pollutants in the area and the levels of individual exposure. Yet the actual extent of the risk will be determined in relation to the established health safety limits; and these have usually been formulated to accommodate to economic and political priorities, rather than to environmental ones. Therefore, while risk assessment can be very useful, it cannot be fully reliable bearing in mind the political agenda; for example, the minimum acceptable level of lead in the blood (that was once thought safe) has been shown to be harmful (Needleman et al., 1979). The established pollutant standards have been conditioned by official acknowledgement of the health risks, and by the will of the government to cut emissions and regulate pollution accordingly. One cannot but conclude that there are intrinsic weaknesses in this legislation. Indeed, had the regulatory system controlled air pollution in absolute accordance with health standards, there would not have been 279 unhealthy and 37 very unhealthy days during the ten-year period between 1981 and 1990 in the Houston area.

Flaws in the air emission regulation

The US regulatory approach is rule-oriented, normally employing rigid and uniform standards, there is executive and judicial scrutiny of regulators, and the use of best available technology rules is favoured. In Houston, most government regulatory work to limit public exposure to toxic air takes the same basic approach. It measures or calculates the concentrations to which the public can be exposed; it reviews information on the effects of each chemical on animals or persons at concentrations that are orders of magnitude higher than the ambient levels; and it determines the regulatory requirements (TACB, 1992e, p. 1). To control and regulate air pollution in Houston, the Texas Clean Air Act (1967) requires the use of best available control technology (BACT) for all new and modified sources of air contaminants; and emissions penalty fees. The rationale behind the use of BACT by the state agency, the TACB, rather than quantitative risk assessment, is that part of the difficulty in setting and using exposure standards comes from the fact that a numerical standard implies that a public exposure at 90 per cent of the standard is acceptable while exposure at 110 per cent of the standard is not. Even if there is a level below which no adverse effects occur, there is a substantial uncertainty in determining that level. Moreover, the state of the science does not allow quantitative risk extrapolation to the exposure levels normally encountered in ambient air.

Typically, regulations which target reduction of toxic emissions do not question the nature of the activities which give rise to air pollution, and this is reflected in the market approach that dominates the field of emissions control. For example, despite Houston being a non-attainment area (Chapter 5), new sources of pollution could be built providing that, first, the proposed plant has installed pollution control technology which ensures the lowest achievable emission rate; and second, that the proposed emissions are offset by reductions in emissions from existing sources in the area (the 'emission offset' policy). The Emissions Trading Policy of 1982 recognizes that air is a scarce resource and uses market forces to accommodate growth without increasing total pollution. Sponsors of a proposed new plant are required to offset the pollution which would result from their plant by reducing emissions from their own plant in the area, by paying another company to reduce its emissions in the area, purchasing an old plant in the area and simply closing it down; or by purchasing emission credits, if available (Elsom, 1992).

This regulatory system allows a degree of flexibility and the reviewer may consider information such as operating schedule, adjacent property, and types of health effects associated with each substance. Also, an emission fee, if applicable, must be paid. The fee for fiscal year 1993 was set at a minimum of $5 per ton, but the fee for fiscal year 1994 and later years is set at a minimum of $25 per ton according to the Texas Air Pollution Regulation (TACB, 1987). Further flexibility is reflected in the 'bubble' strategy introduced in 1979, and incorporated into the Emissions Trading Policy of 1982. The strategy considers the plant, or a series of plants, to be enclosed in a bubble and the EPA sets limits on the total discharges of each type of regulated pollutant, leaving it up to the plant manager to decide how to reach the goals. Trade-offs between sources of pollution within the bubble must involve the same pollutant, so that plants are not allowed to offset, say, reduced sulphur dioxide emission against expensive controls on other toxic emissions (ibid.).

Established industrial plants are controlled in the following way. Any person owning or operating a source of air contaminants must comply with the requirements of EPA for the national ambient air quality standard of the six regulated air pollutants. The TACB states that 'no person shall discharge from any source whatsoever one or more air contaminants or combinations thereof, in such concentration and of such duration as to cause air pollution' (TACB, 1987, p. 14). If any company does so, owners must, upon request by the Board or the executive director, 'conduct sampling to determine the opacity, rate, composition and/or concentration of such emissions' (ibid., p. 15). The TACB establishes that in an area where a cumulative effect occurs from the accretion of air contaminants from two or more sources on a single property or from two or more properties, such that the level of air contaminants exceeds the ambient air quality standards established, and each source or each property is emitting no more than the allowed limit for an air contaminant for a single source or from a single property, 'further reduction of emissions from each source or property will be made' as determined by the Board (ibid., p. 13).

According to the TACB, these procedures have major advantages for the protection of public health because: (1) the BATC requirement minimizes direct public exposure to all chemicals emitted from permitted facilities, whether or not the chemicals are toxic enough to require regulation; and (2) since virtually all chemicals react in the ambient air to form other compounds, some of which are more toxic than the original emissions, minimizing all chemical emissions also minimizes public exposure to the toxic products of atmospheric reactions.

In summary, the use of advanced technology to control industrial pollution and the implementation of numerous regulations have created the false impression that public health enjoys the full protection of the authorities in major cities of the developed world. While these might well be the declared intentions of the regulation, the case study hints at a different reality.

Growing industry and the growth of health risks

After the growth boom of the 1960s and 1970s, and since the depths of a recession in 1987, the city of Houston recorded one of the most dramatic turnarounds in the US, making a rapid economic recovery during the 1990s. However, for decades up until the present, the stench of nearby oil refineries never seemed to concern city leaders: 'The hot summer breezes often carried a brown haze over the skyline and, with it, an odour that the business elite regarded as the sweet smell of money' (Yardley, 2000).

In an era of globalization, unregulated economic growth appears as the natural state of society since current social developments and biophysical change seem unrelated to their past. It is argued that linking current and past growth is necessary to understand the co-evolutionary principle that society, technology, science and the environment are connected. The origins of population settlement in the Houston area date back to the decades of railroad growth in the nineteenth century. Then, the first industrial areas in Houston were those concentrated along the railway, with related machine shops in the northeastern and eastern sectors of the city. A shipping channel, that is, the present Ship Channel which is now the location of the second largest port in the US, was originally dredged in 1857 in the east side of the city to improve poor and shallow port facilities. Residential areas developed in the vicinity of the Ship Channel well before the development of the oil business at the beginning of the twentieth century. They preceded also the rise of petrochemical manufactures in the 1920s, the city's boom of the 1970s, transformation of the city into the unofficial US capital of energy, and considerably antedated the final apotheosis of the city to the oil capital of the world. With the expansion and development of the Ship

Channel in the first decades of the twentieth century new industrial development grew up, running east from the downtown area along the Ship Channel. From the 1920s to the 1940s the eastern side of the city was the location of the vast bulk of oil refining, petrochemical, and metal-related plants and for the developing port facilities.

At each stage of development, whole Houston neighbourhoods were organized around specific types of economic activity (for example, railway workshops, cotton exports, and so on). Despite the development of industrial areas, large sections in the east of the city have remained fully residential until the present day. In fact, new residential development has also taken place in the east. Both detached houses for middle- and upper-income groups and low-standard apartment complexes for the deprived were built in the 1960s–1980s in the east side of the city despite the fact that most suburban development has taken place on the western and northern sides (Feagin, 1988). Contrary to widely held belief, even in the 1990s, residence and industry have remained in close proximity to each other.

A striking feature of this area is air pollution, which cannot but be noticed due to its distinctive sulphuric stench. Often, low white clouds of toxic material emitted by the refineries cover the area and visibility is impaired. The petrochemical plants are certainly located at no 'prudent' distance from residential areas and fumes are more pungent near the sources of emission; none the less, increased distance from the source of emissions has not guaranteed pollution-free neighbourhoods since contaminants reach other parts of the city as well.

Local residents, especially those living in the Ship Channel area, have continually borne the indisputably hazardous burden since oil was first found and extracted in 1901. They are exposed to intense pungent odours from the refinery process, noxious clouds of hydrocarbon gases, and to accidents such as explosions and leakage in petrochemical plants. For example, two massive explosions, in the ARCO Chemical Co. in November 1990, and another in the Phillips Petroleum Co. in October 1989, took the lives of 40 workers and 'showered neighbors with debris' (Grandolfo, 1989, p. 1). Beyond both the initial shock that residents suffer from being in such close proximity to an explosion, and the unknown long-term health effects of the emissions, damage extends to destruction of cars in the vicinity, house foundations, and shattered windows panes (Bardwell, 1989; Pearson, 1990). Indeed, the dangerous effects of such accidents are borne mainly by the local residents:

> Raúl Perez had grown accustomed to the noise and stink from the Houston Ship Channel petrochemical plants. But Monday he heard a distant sound that sent a chill up his spine. 'I was out there when I heard a rumbling at the Phillips plant', he said. 'It wasn't like anything I'd ever heard before, it was eerie. The next thing I knew – boom'. The force of the explosion – several miles away – sent the screen door slamming on his wife and him reeling for cover as a black mushroom cloud rose over the tiny frame houses just north of interstate 10 near Galena Park. Then the rain came. Shards of metal swirled in the dark sky and chunks of pipe insulation hurtled to earth ...
>
> (Grandolfo, 1989, p. A-11)

Apart from oil-related explosions, gas leakage represents a further grave environmental hazard associated with petrochemicals. For example, in September 1990 residents in Deer Park were asked to stay indoors because of a leak of high-pressure gas from the Shell Oil Co. refinery (Byars, 1990). The effects of industrial accidents on the residents have apparently not been taken seriously by the authorities responsible for public health and environmental quality. The only warning that the government issued after a gas leak in 1990 was that residents were at potential danger from skin, eye and throat irritations. Clearly, the risks of more severe short-term reactions such as headaches, sickness and respiratory difficulties are coupled with the possibility of incurring longer-term ill health due to the development of respiratory problems and/or prolonged weakness. Such consequences have gone systematically unacknowledged by the authorities, but these are the real, grim aspects of economic growth and globalization that local residents have to endure.

Being local in a polluted residential area

Despite high concentrations of air pollution in the east side of the city, people live in the surroundings of the Ship Channel. Substantial evidence in the literature, as reviewed in Chapter 2, shows that residential polluted areas are usually occupied by lowest-income populations. The consensus has been that 'where a household resides (and why it resides there) determines the extent to which risks associated with pollution will be experienced' (Murie, 1983, p. 16; see also Townsend, 1988 et al.). In this literature, it is argued from both perspectives, in that areas of bad housing are often those that suffer from other adverse conditions, such as environmental deprivation with no safe place for children to play, for example, and that the highest levels of atmospheric pollution are found in areas with populations of low socioeconomic status (Wood et al. in Murie, 1983).

However, the present research has shown that high socioeconomic status households are also found in environmentally polluted areas. Therefore, it would be mistaken to assume from the outset that families live in the east side of the city of Houston because they cannot afford to live somewhere else, as implied in most of the literature. In a global and growth-oriented city like Houston, low- and high-income households, whether knowingly or not, live in neighbourhoods with considerable levels of air pollution, and one reason is that many activities associated with economic growth have inherently caused widespread environmental contamination. It is the contention here, therefore, that traditional arguments, although perfectly valid, are too narrow. Not only people who live in conditions of poverty have been continually exposed to the externalities of growth; residents on higher incomes have also been severely affected. This has been so because industrial development also occurred in populated areas, state investment expanded into more polluted sectors, and there has been an additional impulse to oil-related activities in this wealthy city. Moreover, certain pollutants expand beyond the proximity of their sources, are broadly dispersed in the air, and combine with other materials to form new pollutants. This situation therefore prompts the question of why people in Houston choose to live in the industrially polluted areas.

In terms of the number of households, 'closeness to work' emerged as the most important factor in the choice of residential area in both the polluted and the less polluted areas (33 per cent and 38 per cent respectively). 'Affordable rents' was the second most important reason for choosing where to live with little variation between the Ship Channel and the Southwest (21 per cent and

20 per cent, poor and rich alike). However, in terms of the most influential factor differentiating between choice of residence in the east of Houston and in the Southwest, 'being local' appeared as the most significant (21 per cent in the polluted Ship Channel compared to only 6 per cent in the Southwest). To 'be local' meant that the respondent had always lived there, and that there was a feeling of belonging to the area. A final factor affecting decisions related to good 'reputation'. 'Reputation' of the area was an important consideration for residents in Houston: good property values (yet house prices were cheaper in the affluent residential Ship Channel than the affluent Southwest), low crime rates, good schools, a friendly neighbourhood, all contributed to the concept 'reputation'. Focusing on high-income households, 44 per cent of respondents in the Ship Channel area and 64 per cent in the Southwest said that they lived there because of the good reputation of the area. Hence neither closeness to nor the stench from refineries has largely altered the good reputation of this particular polluted area.

Old and low-standard detached houses for rent or sale were available and affordable only in the east side of the city. To find cheap detached houses is almost unheard of in the Southwest. In the east, the poor may also live in apartment complexes where rent is generally cheaper than in the Southwest. The only housing facility in the Southwest for low-income households is rented apartments. In the east side of the city, it was easier for the poor to find better accommodation and have better value for rent money. However, the good reputation of the area was equally important for low-income households in the Ship Channel as in the Southwest.

Furthermore, apart from wanting to live in the polluted area because it was close to work, rents were affordable and respondents felt they were 'local', high-income households in the polluted and the less polluted area outlined the importance of kin and friendship networks (33 per cent and 36 per cent respectively). In summary, the reasons residents in the polluted and in the less polluted areas offered in terms of their choice of residential area did not vary significantly with the exception of 'being local'. These findings indicate once more that the east of Houston represents an old, well-established residential area (see Chapter 4), that far from being inhabited principally by a poor population, it contains both those on low and on high incomes, and that the petrochemical industries were only later incomers.

This argument was confirmed by the length of time that residents had stayed in the polluted and in the less polluted areas. It was found that residents in the polluted Ship Channel area settled there earlier and stayed longer than residents in the Southwest (see Figure 7.1), although the fact that the residential Southwest has been more recently developed (since the 1940s) was not the only reason. For example, 34 per cent of all interviewed families in the Ship Channel area had resided there for more than five years, but only 11 per cent had done so in the Southwest. Length of residence for poor households also seemed more stable in the polluted area than in the less polluted area. Indeed, only 24 per cent of low-income households in the polluted, but 50 per cent in the less polluted area lived in the surveyed address for less than one year. The same pattern was found among high-income households: 13 per cent in the polluted and 21 per cent in the less polluted residential area lived for less than one year in the same area.

Having established that 'being local' and 'good reputation' were crucial factors in the choice of residential location in the polluted area indicates that, for the residents, this part of the city was no worse than any other (apart of course from industrial pollution). Perhaps living in the east part of the city was considered a privilege because it was settled earlier than the rest of Houston and represents

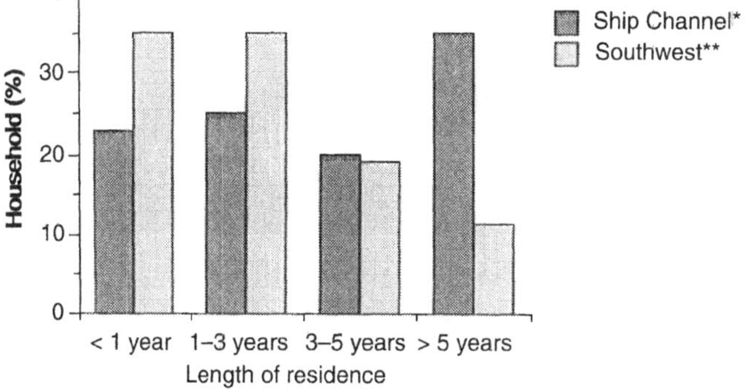

Figure 7.1 Period of time respondents lived in the polluted and in the less polluted residential areas
*Polluted area; **less polluted area

an old-established neighbourhood within a highly respectable world city. The length of time that residents lived in the same area, together with the reasons residents gave for choosing it, suggest that the population here was more established and equally attached to this part of the city. This, in turn, might have had implications for how residents interpreted the problem of local air pollution and of child ill health and their potential power to influence regulatory policy.

In summary, the information from the residents' case study hinted at the fact that industrial growth can take place without any consideration of the hazards it produces (see Plate 5). This participatory or 'lay' knowledge was the expression of the concrete and dynamic characteristics of the contradiction between nature and society under the current conditions of economic growth and

Plate 5 This view shows a household with children in which an impressive petrochemical establishment stands in the background. It illustrates a landscape of hazardous industrial development and residential convergence. It depicts what cannot be directly observed or measured: the disregard that economic growth has attached to the quality of the local surroundings, the negligence of creating pollution, and the imposition of health risks and contamination on residents.

globalization. The 'sweet stench of growth' has been the cost residents had been ready to pay in exchange for consumption: goods, comfort, global power. Yet it is also the case that residents have woken up to the dangers of local contamination and started to question it.

Conclusion

This chapter has implied that neither abundant scientific evidence nor well-intentioned regulations have provided sufficient impetus for change to the growing pollution trends or to protect the city's residents from exposure to them. It must be added, too, that public campaigns by non-governmental organizations such as the American Lung Association, Friends of the Earth and the UK National Asthma Campaign have only had a very limited success in modifying the persistent trend of urban pollution (FoE, 1995; National Asthma Campaign, 1994). In order to explain and regulate contamination and avoid environmental ill health, a proper understanding of the problem is essential.

The chapter has applied a political and economic perspective to emissions regulation. It has been shown that government attempts to assert power in areas previously controlled solely by the corporations met with opposition, particularly over one of the most volatile of the issues, pollution control. Changes in the legislation produced national standards for a number of pollutants and allowed for subsequent federal intervention in state pollution affairs. None the less, neither state nor federal efforts to control air pollution in the Houston area were meant to curtail the tradition of economic growth. It is by now very clear that the outcome of such a tradition has been highly controversial – a controversy reflected in achievements in material and health terms, and in ecological degradation and human costs. The environment has been weakly protected by legislation and, as a result, the population has been exposed to sporadic but severe health risks. Moreover, unequal reporting of ill health in the case study indicates that despite globalization trends and the efforts to enhance the international competitiveness of the city, urban pollution has remained constant and unevenly distributed. Extensive and dangerous air pollution, as it has been unleashed in global cities, shows unequivocally how some of the concrete factors that make the second contradiction of capitalism between society and nature work in the period of globalization.

The previous analysis has indicated that the actual measures implemented to protect citizens from suspected risks caused by development have been minimal, and this despite an evident dedication in the legislation to protect health and the environment. This is a political factor that may help explain in part the positive association between air pollution and ill health in large cities. Whilst the socioeconomic circumstances of the household were significantly associated with the incidence of child ill health, spatial location *vis-à-vis* air pollution sources was highly important in the determination of causality. Indeed, space makes a difference to the events themselves because the conjunction of residential areas, nearby industrial plants, ubiquitous pollution, and substantial ill health has been a concrete manifestation of the way that the environmental regulation has been formulated and has operated in the case of Houston. The particular geographical distribution of child ill health proved the usually discounted human cost of economic growth. Moreover, the spatial analysis also indicated the dynamic, if less visible, existence of social processes. Neither industrialization on its own nor scarce natural resources could explain this important presence, embodied in huge petrochemical facilities, industrial developments and contrasting realities.

The analysis in this chapter has shown that patterns of risk have always existed due to the geographic location of the industrial developments necessary to maintain economic growth. The case study elicited that residents who live in industrially polluted areas consider their neighbourhood no different from any other. Rather, they develop a sense of belonging to the immediate territory, that is, to the geographical area where the household is located. In addition to confirming this important aspect of the relationship between residents and local places, the findings also indicate that the environmental crisis has become the concern of a large number of people, who may or may not live in the most polluted areas of a global city. What this means is that the extent of the problem of hazardous urban environments is perhaps larger and more persistent than previously thought. In a sense, the cause of such problems must be sought beyond the household alone. Explanations based on social inequality, demographic differences, direct exposure to pollutants, deficient access to health care, or inappropriate air control regulations do help, but certainly reduce causality to separate factors. Spatial location of the residence was crucial for identifying how macro-economic and political factors acted in relation to the reported incidence of child ill health. However, it is not spatial location *per se* which accounted for the distribution of child ill health because, as stated in Chapter 2, 'spatial relations are still secondary and contingent, even if primary, generative causal mechanisms are spatially bounded' (Duncan, 1989, p. 135).

The pattern of spatial variation of child illness in Houston reflects the interaction between biophysical and social processes. Significantly, however, the latter operated outside the boundaries of the household. The main feature that made one location different from the other was certainly the proximity of the household to the large petrochemical industry. The case study shows that the effect of spatial location of the house on child health is that of modifying rather than causing the events. Our initial assumption was that the social and the biophysical are interconnected, and that to establish the linkage between the two, wherever possible, more than simple statistical correlation is needed. Also, we try to go beyond the recognition that something produces some change to reach an understanding of what it is about the 'object' of study that enables it to do this (Sayer, 1992, p. 106). This suggests that the value of statistics depreciates as our knowledge of causal mechanisms becomes more complete (Harré, 1970). Yet, statistical methods may still be used to model the relative quantitative dimensions of this group of social and spatial phenomena as long as a qualitative, historical and political interpretation is formed. Pratt has clarified this point by emphasizing that 'groupings of things simply by events do not show any correspondence between causality and event' (Pratt, A.C., 1994, p. 55). The causal factor in this particular location has been the use and abuse of natural resources to promote the growth of the local and national economies.

The information provided by the local respondents has played a very significant role. Particularly affected by industrial pollution have been the socially excluded and the physically vulnerable, people living in decaying neighbourhoods and in areas near sources of industrial emission. However, local high-income residents have also been afflicted by air pollution and the main reason has been the proximity of the residential areas to the industrial plants. This situation could be simply interpreted as the 'democratic' character of pollutants (Beck, 1993; Chapter 2). Yet this particular situation is more eloquent. The environmentally related health of high-income residents has usually been overlooked. Such an oversight can have significant consequences for policy-making. While the prevalence of the illnesses that residents have reported was very similar in poor and wealthy homes, the origin of the incidence was different. For the first type of homes, poverty was the main factor triggering child ill health. In wealthy homes, air pollution was the main cause of the reported ill health. It is not a matter of concluding that it is urgent to reduce emissions because high-income residents are reported to be adversely affected by industrial urban pollution. Rather, the suffering of this social group, which one would expect to be insulated from such consequences, indicates that the gravity of the problem of air pollution in cities is worse than previously thought. Second, ill health in wealthy households indicates that low-income residents bear a double burden: one arising from poverty: the other from bad air. Finally, the oversight will naturally result in an underestimation of the number of the people affected by pollution. The three situations together indicate the need to ensure full consideration of all people affected by pollution, to emphasize the urgent need for more effective policies to control urban air pollution. A primary impetus for these findings has been provided by the weight given to the location of the household in the case study.

Overall urban prosperity and the much-trumpeted integration of local economies into the global business mechanism have apparently obscured, particularly to the layman's eyes, the paradoxical relationship between unrelenting capitalist growth and parallel environmental degradation. The case study has provided invaluable information for uncovering the human face behind the theoretical second contradiction between nature and society as explained in Chapter 1. Long-lasting and politically supported strategies to encourage and maintain economic growth in local places have certainly meant more environmental 'withdrawals' and 'additions'. Local residents have, so far, endured the most direct experience of these phenomena which, in fact, affect us all. In summary, in the face of continuous economic growth and globalization of the economy in wealthy global cities like Houston, neither the protection of environmental quality nor the safety of the population seem to have been a priority. Regulations have been shown to be flexible to the point where these may displace health concerns to accommodate pro-growth activities. Interdisciplinary research has been conducted, but the spatial and the political-economy analyses have been most significant in challenging traditional methods to explain the persistent presence of pollution in large cities. The introduction of US national pollutant standards and state control of industrial emissions at source are very welcome measures to combat air pollution in cities and countries. However, one must appreciate the extent to which these measures were not intended to conflict with political priorities for economic growth. Proper reductions of the allowed levels of industrial emissions can be achieved through more resolute government regulations which communicate less with economic growth and more with the environment and local residents.

8
Conclusion

Introduction

Previous chapters have revealed the enduring character of the problem of air pollution in large cities. Rather than taking a path already heavily trodden by science, the book considered that persistent air pollution in cities represents an uncomfortable enigma for contemporary politics. We have examined the political-economy grounds for current environmental degradation in world cities linking it to past developments in technology and economic policy. Despite successful reductions in the level of certain pollutants, and particularly the improvement in air quality after the 1950s, there has been a miserable failure to reverse the overall rising trend of urban pollution. Increased levels of economic activity lie at the heart of such an unchanging condition. Inspired by Carson, in Chapter 1, we focused special attention on ill health because pollution initiates a downward spiral that destroys not only our world that must support life, but living tissues too, and this damage is mostly irrevocable. Changes in health represent a substantive human indicator associated with future environmental urban sustainability. They are closely linked to ecological and social trends.

The empirical phenomenon of air pollution and economic growth throws greater light on the conceptual contradiction between nature and society in capitalism by further proving that in addition to opportunities, economic growth also brings severe difficulties to the urban environment (O'Connor, J., 1988; Dickens, 1997). The condition of the natural environment and the state of residents' health in a particular global city are useful indicators of the second contradiction, which can best be explained through political-economy theory and interdisciplinary models. The objective of this final chapter is, thus, to bring together the main themes and issues that have emerged from the foregoing discussions and reflections on the pressing biophysical phenomena of air pollution and related ill health. To this end, the politics of environmentalism has provided crucial discrete elements of the theoretical analysis of the combined subject of air pollution and ill health. The internationalization of a regional economy, the social priorities of governments, the experience of the public, geographical distribution of ill health, and the content of the legislation are all numbered among the issues that connect persistent urban pollution in the context of visible economic wealth.

From the comparative assessment of socioeconomic and location factors, it was concluded that ill health was related to outdoor industrial pollution. In world cities, residents can be severely affected by the emissions of long-established oil refineries and other manufacturing plants in the area, by traffic, and by other polluting activities. The geographical convergence of risk factors and actual illness and accidents can best be understood if regulatory and economic factors are taken into consideration. Most importantly, rather than the traditional signposts of poverty and geographical distribution of disease indicating this fact, it was the information gathered from residents in high-income households that illuminated the importance of socially constructed places.

Nearly all approaches to the topic of urban pollution have paid scant attention to what other sciences may say about the same problem from their own different, if nevertheless complementary, perspectives. None the less, the present analysis of air pollution in a global city allows us to deduce more than merely the fact that there has been a crippling fragmentation in the sciences. Three further insights are derived from our investigation. The first judgement relates to the inappropriateness of explaining environmental risks from either a local or a global perspective. A range of localized environmental problems emerges and is sustained by combined local and global economic forces. Also to be considered is the issue of how past and present contamination are linked in post-modern times, for here a political–historical amnesia has prevailed that has blurred this connection. Current policy instruments implemented to reduce industrial toxic emissions are framed to deal with this problem. Finally, there is the question of social, as opposed to individual responsibility for environmental improvement. The combination of pollution and health helps identify the social nature and the relative significance of biophysical as well as social mechanisms by which economic growth proceeds. There is a need to define whose responsibility it is to protect the environment. Particularly, in the light of research that draws on individual aspects of hazardous exposure, on cognitive assumptions that separate the public from the rest of the knowledge, there still remains a separate question. How far should proposed solutions that emerge from the academic literature, political debate and the media assume collective or individualized responsibility for biophysical conditions which are mostly socially determined?

The global nature of local risk

A relatively recent drawback of capitalism has been the globalization of the local, regional and international economies. The mechanics and effects of economic globalization are integral to an understanding of the familiar presence of air pollution in major cities. Gould et al., (1996) claim:

> Socioeconomic contexts which produce ecological problems that are manifest at the local level are increasingly extra-local in the scope of both their economic operations and ecological impacts. As the industrial treadmill expands and matures, and as liquid capital becomes increasingly 'stateless' and borderless, producers are less tied to specific geographic locations for their investment of liquid capital. This provides them with an opportunity to act on a transnational level with unprecedented ease.
>
> (p. 33)

It is clear from the preceding discussion that there is a tight interconnection between the global economy and local levels of air pollution. These links point to two new intricately related processes, the *localization of risks* and the *globality or global nature of local risk*, processes which can be illustrated through the following paradoxical situations.

Persistent local urban pollution has been related, for example, to the manufacturing of energy and to the use of fossil fuels. While these activities occur in particular places, the origin and reach of such operations are purely international (for example, the shipping of raw materials, the export of manufactured products). As the international capital of energy par excellence, Houston epitomizes this paradox. The externalities of these activities invariably exhibit themselves at the local level where they actually take place, a fact which underlines the global origin of an expanding urban *localization of risk* that bears an ongoing and intimate relation to the global economy. One of the ways that local residents experience global growth is through pollution in their own cities. Castells's (1997) views of local territory give us further clues as to how this process happens. He argues that most human experiences and their meanings are still locally based:

> In what is only an apparent contradiction, ecologists are, at the same time, localists and globalists: globalists in the management of time, localists in the defense of space. Evolutionary thinking and policy require a global perspective. People's harmony with their environment starts in their local community.
>
> (p. 127)

On the other hand, globalization has meant that similar activities, patterns of production and consumption, and a large array of cultural effects – such as a uniformity of consumption patterns (outside the scope of this book) – have been reproduced in many places. A consequence has been that urban contamination and associated ill health, such as respiratory problems and the rise in emergency hospital visits, have radiated out in almost every large city, in the developed and developing worlds alike. This is a symmetrical feature of urban environmental degradation, which points towards the other corresponding environmental expression of globalization, that is, the *global nature of local risk*. In the same vein, another issue that points to the escalating interaction between local and global issues is precisely those attempts to distance local pollution from a wider, international political agenda. As discussed throughout this book, air pollution and health have been most often scrutinized as medical and technological problems, often with the addition of individual behavioural and socioeconomic variables (Chapter 2). The issues of health and pollution have together assumed a narrow position within the political environmental agenda that has otherwise criticized the relationship between economic growth and degradation (Chapter 3). This constraint has much to answer for in regard to the exclusion or sidelining of the configuration of urban pollution and ill health from the world of economics and the politics of environmentalism.

A further paradox has emerged. The same local and regional, rather than global, geographical range of the relation between air quality and ill health appears to have contributed to the dismissal of the issue from the political environmental agenda of the last decades. Global environmental problems have become eminently dangerous and affect every corner of this planet. Also, many single local issues have become the responsibility of managing directors of large multinationals, who operate quite elsewhere. Within this widely globalized context, it is apparent that problems concerning the smaller scale, such as urban air pollution, have been left for doctors and nurses to deal with, rather than to politicians. The issue of local air contamination seems to have remained one of low priority for policy-makers, despite the fact that, thanks to technological advances, a well-organized monitoring and public reporting system for daily levels of pollution now exists in all major world cities. It has become imperative to deal with the global environmental crisis, particularly climate change and the loss of biodiversity. Yet local problems surely merit no less attention simply because they lack the apparent drama that accompanies global environmental issues such as rising sea levels and vast holes in the ozone layer. Needless to say, the origin of much of today's air pollution in world cities can be attributed in large measure to the current globalization of the economy.

In summary, in the real world of events, inadequate environmental regulation, rampant economic growth and globalization of local resources have all contributed to transform cities into maelstroms of risk and unpleasantness. Urban air pollution, and related ill health, have been central aspects of this continual explosive mix. A main reason for such a lack of effective regulatory control is that environmental health and pollution are not simply medical or ecological questions, but problems deeply embedded in the social structure, as the localization and global nature of risks have indicated. Concrete geographical variation in child ill health and air pollution is a good example of the way in which equal conditions of globalization in one city can produce different but globally related empirical events. In Chapter 3, the multi-faceted model specified the different research areas that need to be accessed to address this issue. Knowledge of the political and economic structures, health institutions and monitored data could not predict whether, how and where in the city air pollution would affect the population. It is known that particular environmental conditions can provoke asthma attacks and other respiratory diseases. It is, however, only part of the story to acknowledge that either the weather, poverty or individual vulnerability may be the cause of illness. One major benefit of having surveyed the Houston population was therefore the acquisition of this information, impossible to obtain otherwise but essential in order to produce a complete and interdisciplinary explanation based on a political-economy perspective. The human indicator of public and child health in Houston was important in the political and environmental sense (Chapter 6). For one thing, it uncovered subtle flaws in the control of industrial emissions, which could have had an effect on the levels of pollution and also on the state of residents' health (Chapter 7). The other significance of the human indicator was that it provided a first-hand interpretation of the ecological effects of the long-term development of the petrochemical industry within the regional and international economy.

An interdisciplinary political economy

The thrust of this study has been to initiate and prioritize an interdisciplinary approach to the analysis of the various dimensions that compose the study of persistent urban contamination in the context of economic growth. Yearley (1991a) has justly emphasized that 'Although most environmental problems are those of the natural world, and accordingly demand expertise in the natural sciences, this demand is by no means exclusive' (p. 184).

Within the environmental sciences there is an increasing interdisciplinary inclination as well as a parallel tendency to question the adequacy of the traditional sciences for explaining environmental problems. Interdisciplinarity is suggested as the natural normative response which is used in an attempt to expose and supersede disciplinary reductionism in environmental research (Leroy, 1995). This book has attempted to overcome the problem that knowledge in the environmental and health fields assumes an implicit discrete character. The natural and social sciences erect barriers between each other (Chapter 2). A further important aspect to linking distinct sources of information emerges here. The database originating from the collection of local information has generally been used only to explain local situations. This tendency in the literature to confine local studies to local places has reduced the potential of lay knowledge to contribute to a substantial political-economy analysis (e.g., Smith et al., 1999).

The implementation of a voluntary agreement, for example that reached between the US government and the EPA, on the one hand, and the two multinational petrochemical corporations on the other (see below), has required the reduction of industrial emissions and the use of BACT. Local residents' reports of the effects of such modified emissions will certainly provide part of the information necessary to determine whether the level of agreed reductions is the appropriate one, indeed whether it has been a successful legislative instrument to curtail toxic emissions. The two sources, technological and participatory, are necessary to establish a balanced human, political and economic assessment of the development of the industry. In Chapter 7, the assessment of ill health in relation to the location of residential homes left no doubt as to the importance for a political-economy analysis of the data provided by lay residents.

This book was prompted by the urgent need to identify vital connections between ecological biophysical transformations in global cities and the politics of the environment and economic growth. Another objective was to translate conceptual notions, such as the second contradiction of capitalism, cognitive dualism and ahistoricism, into practical guides for research and policy. Linking conceptual and empirical information and generating critical theory is something that may be achieved through a critical realism approach, as explained in Chapter 1. Critical realism has offered a guide to the practice of research and theory generation, which, argues Pratt (1994, p. 204), are interrelated. And theorization has been used as a mode of explanation rather than as an end in itself. This methodology is not a cure for the generalized research limitations of environmental research; it does, however, represent a useful method to elicit a deeper understanding of the contradictions between society and nature (Chapter 3).

Historical amnesia and voluntary agreements

From the study of environmental conditions and ill health in world cities, one may conclude that the ill effects of past episodes of air pollution have apparently been forgotten. It has been argued that the actual significance of current local pollution in ecological and human terms has been sidelined in favour of the general impetus to globalization that dominates the economic climate of the developed world's major cities. The perceived glamour of international competitiveness, higher standards of living (for many but by no means for all), and participation in the global communications network and in worldwide finance have all contributed to the convenience of a generalized 'historical amnesia'. This condition seriously impedes our capacity to see the linkages between past episodes of severe urban contamination and the current less extreme but persistent contamination in world cities.

It is unlikely that current technological advances or scientific research will be able to provide solutions to the dangers of pollution. Despite the fact that a solid body of literature in the biomedical and chemical sciences has pointed unequivocally to the serious consequences of air pollution for health, there is still a large element of uncertainty about the precise dangers posed by old and new industrial developments, new technologies and pollutants released into the environment and how to contain them. The research survey provides a means to bridge some of the gaps between science, technology, social theory and reality. For example, the externalities of growth may well exceed the predictions and descriptions of epidemiological data as described in Chapter 2. Actual events have shown that working refineries, excavation towers, electricity plants, and enormous tanks of gas, petrol and other flammable materials (all of which have been constructed in dangerously close proximity to major population centres) have increased the risk of accidents to local residents (Chapter 7).

The other means to bridge gaps is to incorporate past experience into current reality. The question then arises of how researchers as well as policy-makers might assimilate the experience of the past in interdisciplinary studies of the urban environment. As mentioned in Chapter 3, the co-evolutionary framework has an advantage as an approach to environmental degradation because it provides a model for understanding the way societies have interacted with their environments historically, and suggests new directions for the future. Although co-evolutionism explains the past well, by its nature it does not predict. And yet, perhaps the most important policy lesson to be derived from this understanding of social and environmental process is that

> The belief that we could predict or control environmental outcomes is a delusion. If processes in many cases really are best understood as co-evolutionary, the ability to predict and control will always be limited. And if this is indeed the case, the first policy implication of co-evolutionary environmental sociology is that experimentation should be undertaken frequently, cautiously and on a small scale, with as much monitoring of the evolutionary chain of events thereafter as possible. Massive programmes to quickly adopt new ways of knowing organizing and doing things are inherently risky. Multiple small experiments are better than a few big ones.
>
> (Norgaard, 1997, p. 167)

The principle of adopting small-scale technology experiments in order to safeguard the environment from catastrophic consequences has apparently been applied to policy-making. More out of necessity than choice, the small-scale acts of the US government are aimed at achieving some control over industry in order to protect the environment. Through a combination of voluntary agreement and legal enforcement, the US government has arrived at a record achievement in getting major petrochemical operators/producers to promise a significant reduction in their emissions (US EPA, 2000). Agreements have been signed between the gigantic BP Amoco and the Koch Petroleum Group on the one hand, and the US EPA and the Department of Justice on the other in the year 2000. Considering the continuing failure of cities like Houston to attain the national standards limits for monitored pollutants (Chapter 5), the agreement has justly been considered a major breakthrough in

EPA's enforcement strategy for US refineries under the national 1990 Clean Air Act. The covenant depicts an unprecedented cooperation on the side of the global firms in what is seen as an innovative and comprehensive contract:

> Today's agreement will cut nitrogen dioxide and sulfur dioxide emissions by 49,000 tons annually from the 12 refineries by 2004, and an additional 6,000 tons by 2008, by upgrading the use of new technologies. Improved leak detection and repair practices and other pollution control upgrades will reduce smog-causing volatile organic compounds by 3,600 tons per year and the carcinogen benzene by an estimated 400 tons per year. The agreement also includes measures to improve safety for workers and local communities that will sharply reduce accidental releases of pollutants.
>
> (US EPA, 2000, p. 2)

While the actual effect of reducing emissions as agreed above remains to be seen, voluntary agreements of this type do indicate something else. The hitherto continuous and strong opposition by the industrialist lobby to any environmental control that would affect their business seems to have slightly receded (Chapter 4). It is clear that whatever the implications in the medium and short terms, the population will benefit greatly from an immediate reduction of emissions, even if that implies a slight reduction in the levels of economic activity. The contradiction between nature and society therefore remains at the core of this situation. In fact, the interests of those who run the economy do not always agree with those of society in general. Newby (1991) remin*ds us that significant outcomes* wh*ich result from economic growth* involve, to a greater or lesser extent, the destruction of nature and severe deterioration of human health. Yet the nature and extent of that destruction will reflect the priorities of the society in which that production takes place: 'Beneath the concern for the environment there is, therefore, a much deeper conflict involving fundamental issues about the kind of society we wish to create in the future' (Newby, 1991, p. 2).

Thus, if society's ultimate aim is to achieve growth (historically the prime determinant of economic and social decision-making), rather than to have a profound commitment, say, to improving the general standard of the population's health, or attaining low levels of particulate matter and ozone everywhere, this will be reflected in current and future patterns of environmental quality and of health status in cities. The view that human progress can only be attained through economic growth of the type pursued in the last 50 years, which disregards environmental and human health costs, has been challenged in this book. Raising this concern now is vital because a prospective reduction in noxious emissions as a result of agreements such as the one previously mentioned may well require a reduction in the levels of economic growth which will undoubtedly encounter opposition. It is likely that a complex effort will be needed to maintain the voluntary agreement policy. Such a reduction in the level of industrial activity can, however, be easily justified on the grounds of significant improvement in human health.

The trend of ever-increasing air pollution in cities needs to be controlled and reversed. Many factors furnish sufficient evidence of malign consequences for us to want to overcome the historical amnesia that characterizes our contemporary globalized post-modernity. To mention but two: the infamous results of the air pollution episodes that occurred over Europe in the 1990s; and the fact that the US is one of the nations with the highest prevalence of asthma (Chapter 1). These findings attest the fact that air pollution in industrialized cities causes widespread respiratory illness and also death; that one out of five children in the UK suffers from asthma, and that mortality rates in certain cities are raised due to air pollution (Chapter 2). Such realities indicate that it is up to governments, the research community, international firms, other stakeholders, and the public alike to work for the emergence of the most appropriate and stringent pollution control rules in highly developed cities.

The responsibility for environmental protection

The state of the environment in global cities must be understood as a complex phenomenon, for explanatory as well as policy purposes. A reductionist interpretation only encourages further environmental degradation. Various dimensions cross-cut the subject; none could be assumed to determine the other in analytical terms. It has also been argued that dualistic thought was not helpful for an understanding of the presence of pollution in cities. Any single choice of focus, for example on social structures and globalization, on the epidemiology of environmental health in Houston, on information obtained from local residents, or on the numbers produced by monitoring instruments would have mirrored the typical dualities between the social and natural sciences, between expert and lay knowledge. It emerges, however, that dualistic philosophical assumptions of the type described in Chapters 1 and 2 are at the core of pragmatic implications about who must be made responsible for improving the quality of the urban environment.

The course of rising environmental degradation in large cities has often been narrowly understood, and often sidelined, during the last decades. It has been argued that trends such as increased production, expanding consumption, globalization of cities and urban competitiveness have served to obscure the relevance of the problem of urban contamination as a local and also global phenomenon (Chapter 4; see above). The population at large has hardly questioned these economic trends. Unless city-dwellers have been directly confronted by overt environmental health risks such as those associated with the international transportation of toxic waste, few voices have been raised against the environmental costs of economic growth. The case of Houston illustrates this point. Residents in this highly economically developed city had been used to 'the sweet stench of growth', that is, to the odour typical of the city odours originating in a successful petrochemical and refinery oil business, heavy traffic, and other industrial activities. The smell, in reality, is of air filled with filthy material – atmospheric ozone, sulphur dioxide, particles and volatile organic compounds emitted by the sources of much wealth (Chapter 4). Breathing in toxic cocktails has been a high cost that residents have paid for uncontrolled economic growth and weak environmental legislation. The data from the environmental surveys, the HAS and the TES in Chapter 5, and the APCHS in Chapter 7, indicated that local residents are no longer as ready as they were previously to endure local chemical pollution. Despite the active role residents can play, it is not the responsibility of individual residents to reverse urban pollution trends.

With this scenario in mind, let us now turn to examine how it is that the responsibility for improving the quality of the environment has been left by default to individuals rather than to social structures. This individualization springs from the scientific literature in particular. While an underlying tenet of the natural sciences is to focus on the individual, Rose et al. (1984) expressed a deep antipathy for a philosophical tradition of individualism which emphasized the priority of the individual over the collective. Such a view invokes a sense of inevitability because what is given by nature and shown by science is claimed to be unchangeable. The political implications of individualism are that by understanding individual behaviour it becomes possible to understand society. Accordingly, the properties of a human society are, similarly, no more than the sums of the individual behaviours and tendencies of the individual humans of whom that society is composed. Following this assumption, it would be impossible to create a different society from the one that is considered to be 'natural'. An individualist, deterministic philosophy ignores the fact that society is an open system with structures and emergent powers, with influential geographical and local factors, and with significant links to the global economy (Chapter 3). A dualism between the social and natural sciences has contributed to mythologizing the links between environment and politics (Redclift, 1984). This separation has been influential in assigning who must be responsible for environmental degradation. For all practical purposes, the particular variations of the findings of the case study in Houston (Chapter 5) indicate that, while the effects of pollution and other factors can be assessed by individual households, political institutions are the main institutions responsible for the presence of hazardous materials, increased risk, and also, in the last resort, for dealing with them.

Recognition of the possibility that greater and lesser levels of pollution and ill health may often coexist within the same large and wealthy city does little to alleviate this problem. The fact that children in Houston were differently affected by air pollution (an outcome which depends on location and socioeconomic conditions) in no way neutralizes the adverse effects of the acute contradiction between economic growth and environmental degradation. The origin of urban contamination, as well as the factors that affect individual ill health, remain with the organization of society and its mode of production which, 'naturally', under current conditions of globalization and growth, promote prosperity for the few and long-term environmental and social misery for the many.

The public has a crucial role to play by expressing their disquiet, describing their experience and making explicit requests about priorities. This is one way forward for the public at large to redress their previous social and political exclusion in environmental matters. It is also a crucial way for society to challenge the pressing contradictions between economic growth and environmental degradation. However, it is profoundly inappropriate to make residents rather than society responsible for the changes needed to protect the environment. This book has not offered a blueprint for policy initiatives that could reduce urban contamination, nor suggested guidelines for local organization around pollution issues. Yet, applying the insights discussed in this last chapter will have implications for defining whose social responsibility it is to care for the environment and residents' health, and what type of change is desirable. The task of this book has been to indicate a way to link the biophysical events of air pollution in cities with the political and economic mesh that permits them. It is suggested that to reverse urban contamination, governments in particular must analyse the relevant episodes from the past to guide future policy.

Appendix The Household Survey and List of Interviewees

The Air Pollution and Child Health Survey (APCHS)

The Household Questionnaire

The focus of data collection on the state of child health and local environmental conditions was the Houston household questionnaire. The term *household* has been used according to the definition of *The General Household Survey*, 1971. A *household* unit consists of members of one household who are a couple or one person without a partner and any of their children, provided these children have never themselves been married and have no children of their own (Office of Population Censuses and Surveys, Central Statistical Office, 1973).

The questionnaire produced a large amount of data on 300 households in Houston. However, it also yielded useful and relevant information to determine causal relationships because the socioeconomic conditions and location of the household were formerly identified. The questionnaire was thus used to collect standardized information from a randomly selected population, i.e., in low- and high-income households located near to and far from (used primarily as the control population) the petrochemical and industrial area and with high and lower levels of air pollution which enabled the author to carry out cross-sectional analysis for comparative study (Vaus, 1986; Healey, 1990; Robson, 1993).

The household questionnaire, The Air Pollution and Child Health Survey – the APCHS, Houston, 1990 – was conducted between June and October 1990 in the Ship Channel and the Southwest areas. A closed format was used for the question design, with a number of alternative answers for most questions. Although the questionnaire was based on structured interviewing, by administering it personally it was possible to gather additional comments by the respondents. Only households with at least one child were surveyed and only one questionnaire was filled in for each household. Interviewing was carried out by the author. The questionnaire was addressed to the mother because of her active economic and social role in the household, because of the large number of single-parent (mother) households in Houston (and in the sample), and because mothers are usually more involved with the health of children. In a few instances, when the mother was not at home, fathers answered the questionnaire. The questionnaire was administered either in English or in Spanish (it was anticipated and found that many respondents, particularly in the low-income areas, would be Hispanic). The response rate was 100 per cent although sometimes certain questions in the questionnaire were not answered.

The selection of households and geographical areas

A representative sample in each of the polluted and the less polluted areas of 150 households was interviewed with half being low- and half high-income households. Each household on a selected street had an equal chance of being interviewed. The strategy for household selection was quasi-random sampling because it was essential that there would be at least one child in a household and this was not known beforehand. If there was no child in the household, then another household in the same street was added to the survey until the 300 household interviews were obtained.

The sample areas within which the household survey was conducted in Houston were randomly selected and the selection was heavily influenced by three factors: the socioeconomic features of the residential areas; their location in relation to industrial sources of air pollution; and their levels of measured air pollution. Representative households within the city were thus selected by way of cluster sample, which involved the random selection of two geographical units combined with the assessment of every case within the geographical units. Therefore, the selection of two geographical areas responded to locational (when this was related to economic activity in the city), ecological and social characteristics: one near and the other far from the petrochemical complex and port facilities in the Ship Channel (particularly within the boundaries of two city Health Service Areas, Magnolia and the Southwest), one with highly polluted air and the other with considerably less polluted air; and both having poor and wealthy households.

The information initially used to identify social characteristics of different geographical areas in Houston was elicited from the summary of population profiles of Houston public health annual reports published by the City of Houston Health and Human Services Department (1984–88). This was, however, insufficient and inadequately qualified. Levels of air pollution in different sectors of Houston were examined for the years 1989–90 using the BAQC air pollution monthly reports, particularly for ozone and sulphur dioxide, in two distant stations, in the 9525 Clinton Drive site in the Ship Channel and at 13826 Croquet Drive in the Southwest. Information on the levels of air pollution in Houston was expanded significantly in later stages of the research by examination of the State of Texas TACB and the US EPA publications on air pollution. The existence of low- as well as high-income households was verified by observing the external character and size of the dwellings, the state of outside gardens, whether the streets were paved, types of cars parked in driveways, and whether there were trees on the pavement and shops in the area.

Quantitative analysis

Univariate, bivariate and multivariate analyses were carried out and nominal and interval variables were used. Analysis of frequency was initially performed to describe the composition of the household and the extent of child ill health. The statistical functions analysed were cross-tabulations, correlations and regressions. Tables and graphs were used to display the frequencies and the relationships.

The main correlation coefficient used in the tables and figures was the chi-square. It was employed to evaluate the probability of an apparent difference occurring by chance between the four groups of households in the low- and high-income brackets in polluted and less polluted residential areas. The 'P-value' shown in tables and figures is the only inferential function used in the thesis; it is not presented as establishing causality but is the probability of the particular difference occurring by chance (The P-value is shown only when it was less than 0.05, that is, when the relationship was statistically significant). Standard deviation is used in frequencies (Healey, 1990). The statistical package used was SPSS/VAX. Simple descriptive and inferential statistical analyses were carried out (Healey, 1990; Norusis, 1988, SPSS Inc., 1989).

List of officials interviewed

- Ada Montalvo, Responsible for Casa María Volunteers Clinic, Southwest Houston, July 1990
- Dr Aaron Mintz, Head of Paediatrics, Ben Taub County Hospital, February 1992
- Dr Steven Klineberg, Department of Sociology, Rice University, Houston; author of the first environmental survey in Houston, February 1992
- Dr Sulabaha Hardikar, City of Houson Health and Human Services Department, February 1992.
- Dr Virginia Moyer, Community Health, Paediatrics, L.B. Johnson County Hospital, February 1992; telephone interview
- Gene McMullen, Director, City of Houston Air Quality Control Bureau, February 1992

Bibliography

Adam, B. (1994), 'Running out of time: global crisis and human engagement', in M. Redclift and T. Benton (eds), *Social Theory and the Global Environment*, London and New York: Routledge.

Adams, J. (1995), *Risk*, London: UCL Press.

Agency for Toxic Substances and Disease Registry (1988), *The Nature and Extent of Lead Poisoning in Children in the United States: A Report to the Congress* Atlanta, Ga: US Department of Health and Human Services.

Aligne, A.C., Auinger, P., Byrd, R.S. and Weitzman, M. (2000), 'Risk factors for pediatric asthma. Contributions of poverty, race and urban residence', *American Journal of Respiratory and Critical Care Medicine*, 162, 873–7.

American Lung Association (1994), *Lung Disease Data 1994*, New York: ALA.

Anderson, H.R., Spix, C., Medina, S., Schouten, J.P., Castellsague, J., and Rossi, G. (1997), 'Air Pollution and Daily Admissions for Chronic Obstructive Pulmonary Disease in Six European Cities', *European Respiratory Journal*, 10; 1064–71.

Annette, P., Tuch, T., Brand, P., Heyder, J. and Wichman, H. E. (1996), 'Size distribution of ambient particles and its relevance to human health', *Proceedings of the Second Colloquium on Particulate Air Pollution and Human Health*, 1–3 May Utah, 4-406–4-412.

Arsen, J. and Darnay, D. (eds) (1994), *Statistical Record of Health and Medicine* Detroit, US:Code and Data Inc. Staff.

Ashton, J. (ed.) (1993), *Healthy Cities*, Milton Keynes and Philadelphia, PA: Open University Press.

Ashton, J. and Ubido, J. (1991), 'The healthy city and the ecological idea', *The Society for the Social History of Medicine*, pp. 171–80.

Bach, W. (1972), *Atmospheric Pollution*, New York: McGraw-Hill.

Badham, R.J.(1982), *Theories of Industrial Society*, London: Croom Helm.

Baghurst, P.A., McMichael, A.J., Wigg, N.R., Vimpani, G.V., Robertson, E.F., Roberts, R.J. and Tong, S.L. (1992), 'Environmental exposure to lead and children's intelligence at the age of seven years', *New England Journal of Medicine*, **327**, (18), 1279–84.

Bahro, R. (1986), *Building the Green Movement*, London: GMP.

Ball, S. and Bell, S. (1994), *Environmental Law*, 2nd edn. London: Blackstone Press.

BAQC, *1988 City of Houston Annual Data Report*, Houston, TX: City of Houston Health and Human Services Department.

BAQC, *1989 City of Houston Annual Data Report*, Houston, TX: City of Houston Health and Human Services Department.

BAQC, *1990 City of Houston Annual Data Report*, Houston, TX: City of Houston Health and Human Services Department.

BAQC, (1991), *1991 City of Houston Annual Data Report,* City of Houston Health and Human Services Department: Houston, TX.

BAQC, *1992 City of Houston Annual Data Report,* Houston, TX: City of Houston Health and Human Services Department.

Bardwell, S.K. (1989), 'Ball of fire – then panic', *The Houston Post,* 24 October, p. 1.

Bartelmus, P. (1986), *Environment and Development,* London: Allen and Unwin.

Bartelmus, P. (1994), *Environment, Growth and Development. The Concepts and Strategies of Sustainability,* London and New York: Routledge.

Bascom, R., Bromberg, P.A., Costa, D.A., Devlin, R., Dockery, D.W., Frampton, M.W., Lambert, W., Samet, J.M., Speizer, F.E. and Utell, M. (1996), 'State of the Art. Health Effects of Outdoor Air Pollution – Part 1', *American Journal of Respiratory and Critical Care Medicine,* 153, 3–50.

Beck, U. (1993), *Risk Society. Towards a New Modernity,* 2nd edn. New York: Sage.

Beck, U. (1997), 'Global risk politics', in Michael Jacobs (ed.), *Greening the Millennium? The New Politics of the Environment,* Oxford: Blackwell.

Benton, T. (1989), 'Marxism and natural limits: an ecological critique and reconstruction', *New Left Review,* 178, 51–86.

Benton, T. (1991), 'Biology and social science: why the return of the repressed should be given a (cautious) welcome', *Sociology,* **25** (1), 1–29.

Benton, T. (1994), 'Biology and social theory in the environmental debate, in M. Redclift and T. Benton (eds), *Social Theory and the Global Environment,* London and New York: Routledge.

Benton, T. and Redclift, M. (1994), 'Introduction', in M. Redclift and T. Benton (eds), *Social Theory and the Global Environment,* London and New York: Routledge.

Bertell, R. (1985), *No Immediate Danger. Prognosis for a Radioactive Earth,* London: The Women's Press.

Bhaskar, R. (1975), *A Realist Theory,* Leeds: Leeds Books.

Bhaskar, R. (1979), *The Possibility of Naturalism,* Brighton: Harvester.

Black, Sir D., Morris, J.N., Smith, C. and Townsend, P. (1982), *The Black Report. Inequalities in Health,* Harmondsworth, UK: Penguin Books.

Blackburn, C. and Graham, H. (1992), *Smoking Among Working Class Mothers,* University of Warwick, UK: Department of Applied Social Sciences.

Blackman, T., Evason, E., Melaughs, M. and Woods, R. (1989), 'Housing and health: a case study of two areas in West Belfast', *Journal of Social Policy,* **18** (1), 1–26.

Blaikie, P. (1985), *The Political Economy of Soil Erosion in Developing Countries,* New York: Longman Scientific and Technical.

Bland, J.M., Holland, W.W. and Elliot, A. (1974), 'The development of respiratory symptoms in a cohort of Kent school children', *Bulletin de Physio-Pathologie Respiratoire (Nancy),* 10, 699–715.

Blaxter, M. (1975), 'Social class and health inequalities', in C.O. Carter and J. Peel (eds), *Equalities and Inequalities in Health,* London: Academic Press.

Blaxter, M. (1981), *The Health of Children. A Review of Research on the Place of Health in Cycles of Disadvantage,* London: Heinemann Educational.

Blaxter, M. (1983), 'Health services as a defence against the consequences of poverty in industrialised societies', *Social Science and Medicine*, **17** (16), 1139–48.

Blaxter, M. (1990), *Health and Lifestyles*, London and New York: Routledge.

Blowers, A. (1984), *Something in the Air. Corporate Power and the Environment*, London: Harper and Row.

Blowers, A. (1989), 'Radioactive waste – local authorities and the public interest', *Radioactive Waste Management*, British Nuclear Energy Society, 2, 26–7.

Blowers, A. (1993), 'Environmental policy: the quest for sustainable development', *Urban Studies*, **30** (4/5), 775–96.

Blowers, A. and Lowry, D. (1987), 'Out of Sight, Out of Mind: the politics of nuclear waste in the United Kingdom', in A. Blowers and D. Pepper (eds), *Nuclear Power in Crisis*, London: Croom Helm, pp. 129–63.

Bobak, M. and Leon, D. (1992), 'Air pollution and infant mortality in the Czech Republic, 1986–1988', *Epidemiology*, 340; 24 October, 1010–14.

Boisseau, C. (1990), 'Richest of All', *Houston Chronicle*, 9 October, p. 1C.

Boone, M.S. (1989), *Capital Crime. Black Infant Mortality in America*, New York: Sage.

Boussin, G., Cayla, F., Giroux, M., Carrière, I., Fernet, P. and Pous, J. (1989), 'Pollution atmosphérique et pathologie respiratoire aigüe de 1000 enfants à Toulouse', *Pollution Atmosphérique*, October–December, 387–96.

Brace, M. (1994), 'London air worst for 40 years', *The Independent*, 30 June, p. 4.

Bravo Alvarez, H. (1987), *La Contaminación del Aire en México*, México City: Universo Veintiuno.

Breheny, M.J. (ed.) (1992), *Sustainable Development and Urban Form*, London: Pion.

Brimblecombe, P. (1988), *The Big Smoke: a History of Air Pollution in London Since Medieval Times*, 2nd edn. London: Methuen.

Britton, M., Fox, A.J., Goldblatt, P., Jones, D.R. and Rosato, M. (1990), 'The influence of socio-economic and environmental factors on geographic variations in mortality in OPCS', *Mortality and Geography*, London: HMSO.

Brown, L.R., Flavin, C. and Pastel, S. (1989), 'A world at risk', in L.R. Brown, *State of the World. A Worldwatch Institute Report on Progress Toward a Sustainable Society. 1989*, New York and London: W.W. Norton.

Brown, M.J., DeGiacomo, J.M., Gallagher, G., Graef, J., Leff, J., Mathieu, O., Petre, R. and Sagov, S. (1990), 'Lead poisoning in children of different ages. Letter to the editor', *The New England Journal of Medicine*, **323** (2), July, 135–6.

Brown, P. (1991), 'Stop driving plea for polluted London', *The Guardian*, 14 December, p. 1.

Brown, P. (1992), 'Air pollution putting health of 1 in 3 at risk', *The Guardian*, 27 July, p. 3.

Brunekreef, B.(1997), 'Air Pollution and Life Expectancy: Is There a Relation?', *Occupational and Environmental Medicine*, 54; 781–4.

Burguess, R.G. (1991), *In the Field. An Introduction to Field Research*, London: Allen and Unwin.

Burr, M.L., Verrall, C. and Kaur, B. (1997), 'Social deprivation and asthma', *Respiratory Medicine*, 91 (10), 603–8.

Butler, N. and Golding, J. (1986), *From Birth to Five: a Study of the Health and Behaviour of Britain's Five Year Olds*, Oxford: Pergamon Press.

Buttel, F.H. (1997), 'Social institutions and environmental change', in Michael Redclift and Graham Woodgate (eds) *The International Handbook of Environmental Sociology* Cheltenham, UK and Brookfield, US: Edward Elgar.

Byars, C. (1990), 'Portion of highway closed by gas leak', *Houston Chronicle*, 30 September, p. 13.

Caldwell, M. (1977), *The Wealth of Some Nations*, London: Zed Press.

Capra, F. (1983), *The Turning Point: Science, Society and the Rising Culture*, London: Fontana.

Caprio, R., Margulis, H. and Joselow, M. (1975), 'Residential location, ambient air lead pollution and lead absorption in children', *The Professional Geographer*, **XXVII** (1), February, 37–41.

Carson, R. (1962), *Silent Spring*, Boston: Houghton Mifflin.

Castells, M. (1997), *The Power of Identity Volume II*, Oxford: Blackwell.

Castells, M. (1998), *End of Millennium. The Information Age. Economy, Society and Culture. Volume III*, Oxford: Blackwell.

Castellsague, J., Sunyer, J., Sáez, M. and Anto, J.M. (1995), 'Short-term association between air pollution and emergency room visits for asthma in Barcelona", *Thorax*, 50, 1051–56.

Centers for Disease Control (1988), 'Childhood Lead poisoning – United States: Report to the Congress by the Agency for Toxic Substances and Disease Registry', *Mortality and Morbidity Weekly Report*, 19 August, 37 (32), 481–5.

Chapman, K. and Walker, D. (1987), *Industrial Location*, Oxford: Basil Blackwell.

Cherni, J.A. (1992), 'Unseen costs of development: industrial air pollution and child health variation: the case of Houston, USA', in C. McIlwaine and G. Ozcan (eds), *Geography Discussion Papers. Environmental Debates*, New Series No. 26, London: Graduate School of Geography, London School of Economics and Political Science.

Cherni, J.A. (1993a), 'Child health, air pollution and political economy in the Texan oil city of Houston', in T. Driver and G. Chapman (eds), *South–North Centre Series. Political Economy and the Environment*, 1: 4, University of London.

Cherni, J.A.(1993b), 'Urban environmental pollution and child health in Houston, USA: the links to economic growth', in J. Holder, P. Lane, S. Eden, R. Reeve, U. Collier and K. Anderson (eds), *Perspectives on the Environment*, Aldershot, UK: Avebury.

Child Hill, R. and Feagin, J.R. (1987), 'Detroit and Houston: Two cities in global perspective', in J.R. Feagin and M.P. Smith (eds), *The Capitalist City. Global Restructuring and Community Politics*, Oxford: Basil Blackwell.

Children at Risk Committee (1990), *Children at Risk*, Houston: March of Dimes.

Children's Defense Fund (1990), *Children First*, Texas, Washington DC.

City of Houston Health and Human Services Department (1989a), *The Health of Houston 1984–1988. I.*

City of Houston Health and Human Services Department (1989b), *The Health of Houston 1984–1988. II.*

City of Houston Health and Human Services Department (1990a), *The Health of Houston 1985–1989. I.*

City of Houston Health and Human Services Department (1990b), *The Health of Houston, 1985–1989. II.*

City of Houston Health and Human Services Department (1991), *The Health of Houston, 1989–1990.*

City of Houston Health and Human Services Department (1992), *The Health of Houston, 1991.*

Clapp, B.W. (1994), *An Environmental History of Britain since the Revolution*, London: Longman.

Cloke, P., Philo, C. and Sadler, D. (1991), *Approaching Human Geography*, London: Paul Chapman.

Cohen, M.J. (1997), 'Risk society and ecological modernisation: alternative visions of post-industrial nations', *Futures*, 29, 105–19.

Cohen, R.B. (1981), 'The new international division of labour, multinational corporations and urban hierarchy', in M. Dear and A.J. Scott (eds), *Urbanization and Urban Planning in Capitalist Society*, London: Methuen.

Colección Medio Ambiente (1992), *México City*, México City: Universo Veintiuno.

Colley, J.R. and Reid, D.D. (1970), 'Urban and social origins of childhood bronchitis in England and Wales', *British Medical Journal*, 2, 213–19.

Colley, J.R., Douglas, J.W.B. and Reid, D.D. (1973), 'Respiratory disease in young adults: Influence of early childhood lower respiratory tract illness, social class, air pollution, and smoking', *British Medical Journal*, 3, 1031–34.

Collins, J.J., Kasap, H.S. and Holland, W.W. (1971), 'Environmental factors in child mortality in England and Wales', *American Journal of Epidemiology*, 93, 10–22.

Comisión Metropolitana (1992), *Qué Estamos Haciendo para Combatir la Contaminación del Aire en el Valle de México?*, Comisión Metropolitana para la prevención y control de la contaminación ambiental en el valle de México.

Commoner, B. (1972), *The Closing Circle*, London: Jonathan Cape.

Connor, S. (1994), 'Health advisers meet in secret over deadly smog of 1991', *The Independent*, 26 June, p. 4.

Conzen, M.P. (ed.) (1994), *The Making of the American Landscape*, New York and London: Routledge.

Cox, K.R. (ed.) (1997), *Spaces of Globalization: Reasserting the Power of the Local* New York: UCL Press.

Crilly, M., Morris, A. and Marrow (1999), 'Indicators for change: taking a lead', *Local Environment*, **4**, 2, 151–68.

Cuzick, J. and Elliot, P. (1992), 'Small-area studies: purpose and methods', in P. Elliot, J. Cuzick, D. English and R. Stern (eds), *Geographical and Environmental Epidemiology: Methods for Small Areas Studies*, Oxford: Oxford University Press, pp. 10–21.

Daly, H. (1992), *Steady-State Economics*, 2nd edition, London: Earthscan.

Davies, J.K. and Kelly, M.P. (1993), *Healthy Cities. Research and Practice*, London and New York: Routledge.

Davies, R. (1994), 'What exactly does vehicle pollution do to the lungs?', in *How Vehicle Pollution Affects Our Health*, London: The Ashden Trust.

Dawson, B. (1990), 'Poll: Texans want tough pollution laws', *Houston Chronicle*, 2 October, p. A-9.

Dawson, B., Horobin, G., Illsley, R. and Mitchell, R. (1969), 'A survey of childhood asthma in Aberdeen', *Lancet*, 1, 826–31.

Delfino, R.J., Becklake, M.R. and Hanley, J. (1994), 'The relationship of urgent hospital admissions for respiratory illnesses to photochemical air pollution levels in Montreal', *Environmental Research*, 67, 1–19.

Department of the Environment, The Quality of Urban Air Review Group (1992), *Urban Air Quality in the United Kingdom*, London: Air Quality Division, Department of the Environment.

Department of the Environment (1994a), *Expert Panel on Air Quality Standards. Benzene*, London: Air Quality Division, Department of the Environment.

Department of the Environment (1994b), *Improving Air Quality*, London: Air Quality Division, Department of the Environment.

Department of Health, Advisory Group on the Medical Aspects of Air Pollution Episodes (1995b), *Health Effects of Exposures to Mixtures of Air Pollutants*, London: HMSO.

Department of Health, Committee on the Medical Effects of Air Pollutants (1995a), *Asthma and Outdoor Air Polllution*, London: HMSO.

Dickens, P. (1992), *Society and Nature. Towards a Green Social Theory*, New York and London: Harvester Wheatsheaf.

Dickens, P. (1997), 'Beyond sociology: Marxism and the environment', in Michael Redclift and Graham Woodgate (eds) *The Environmental Handbook of Environmental Sociology*, Cheltenham, UK and Brookfield, US: Edward Elgar.

Dickens, P., Duncan, S., Goodwin, M. and Gray, F. (1985), *Housing, States and Localities*, London and New York: Methuen.

Dobson, A. (ed.) (1991), *The Green Reader*, London: André Deutsch.

Dobson, A. (1995), *Green Political Thought*, 2nd edn. London and New York: Routledge.

Dockery, D.W, Ware, J.H., Ferris, B.G., Speizer, F.E., Cook, N.R. and Herman, S.M. (1977), 'Change in pulmonary function in children associated with air pollution episodes', *Journal of Air Pollution Control Association*, 32, 937–42.

Dockery, D.W., Speizer, E., Stram, D.O., Ware, J.H., Spengler, J.D. and Ferris, B.G. (1989), 'Effects of inhalable particles on respiratory health of children', *American Review of Respiratory Diseases*, 139, 587–94.

Dockery, D.W., Pope III, C.A., Xu, X., Spengler, J.D., Ware, J.H., Fay, M.E., Ferris, B.G. and Speizer, F.E. (1993), 'An association between air pollution and mortality in six U.S. cities', *The New England Journal of Medicine*, **329** (24), 1753–59.

Dorning, M. (1990), 'Concern for environment has yet to peak, survey finds', *Houston Post*, 21 March, p. A-12.

Dougherty, C.J. (1985), *Ideal, Fact and Medicine*, Lanham, MD: University Press of America.

Dougherty, C.J. (1988), *American Health Care. Realities, Rights and Reforms*, Oxford: Oxford University Press.

Dougherty, G.E. (1986), *'Socioeconomic differences in pediatric mortality in urban Canada*: 1981', Department of Epidemiology and Biostatistics, Montreal: McGill University Press.

Douglas, M. and Wildavsky, A. (1983), *Risk and Culture: An Essay on the Selection of Technological and Environmental Dangers*, Berkeley, CA: University of California Press.

Doyal, L. (1987), *The Political Economy of Health*, London: Pluto Press.

Draper, P. (ed.) (1991), *Health Through Public Policy. The Greening of Public Policy*, London: Green Print.

Duhl, L. (1986), 'The Healthy City: Its function and its future', *Health Promotion*, I, 55–60.

Duncan, S.S. (1989), 'Uneven development and the difference that space makes', *Geoforum*, **20** (2), 131–9.

Duncan, S.S. (1991), 'The geography of gender divisions of labour in Britain', *Transactions of the Institute of British Geography*, 16, 420–39.

Duncan, S.S. and Goodwin, M. (1987), *The Local State and Uneven Development. Behind the Local Government*, New York: St Martin's Press – now Palgrave.

Edgar, D., Keane, D. and McDonald, P. (1989), *Child Poverty*, Sydney: Allen and Unwin.

Ehrlich, P.R. and Ehrlich, A.H. (1970), *Population Resources Environment: Issues in Human Ecology*, San Francisco, CA: W.H. Freeman.

Elkin, T. and McLaren, D. (1991), *Reviving the City. Towards Sustainable Urban Development*, London: Friends of the Earth.

Elkin, T. and McLaren, D. with Hillman, M. (1991), *Reviving the City. Toward Sustainable Urban Development*, London: Friends of the Earth.

Elkington, J. (1987), *The Green Capitalists*, London: Victor Gollancz.

Elkington, J. and Hailes, J. (1988), *The Green Consumer Guide*, London: Victor Gollancz.

Elkins, P. (1993), '"Limits to growth" and "sustainable development": grappling with ecological realities', *Ecological Economics*, **8**(3), 269–88.

Elsom, D.M. (1992), *Atmospheric Pollution. A Global Problem*, 2nd edn. Oxford: Basil Blackwell.

Elsom, D.M. (1996), *Smog Alert. Managing Urban Air Quality*, London: Earthscan.

EMEP, European Monitoring and Evaluation Programme (1997), *Transboundary Air Pollution in Europe*. MSC/W Report 1/1997, Oslo: EMEP.

ENDS (Environmental Data Services) (1994a), 'Sharp rise in asthma cases linked to air pollution', *ENDS Report*, 231, April, London: ENDS.

ENDS (Environmental Data Services) (1994b), 'Ozone pollution review', *ENDS Report*, 231, April, London: ENDS.

Eyles, J. and Woods, K.J. (1983), *The Social Geography of Medicine and Health*, London and Canberra: Croom Helm.

Feagin, J.R. (1985), 'The global context of metropolitan growth: Houston and the oil industry', *American Journal of Sociology*, **90** (6), 1204–30.

Feagin, J.R. (1987), 'The secondary circuit of capital: office construction in Houston, Texas', *International Journal of Urban and Regional Research*, **11**(2), 172–93.

Feagin, J.R. (1988), *Free Enterprise City: Houston in Political-Economic Perspective*, New Brunswick, NJ and London: Rutgers University Press.
Feagin, J.R. and Smith, M.P. (1987), 'Cities and the new international division of labor: an overview', in J.R. Feagin and M.P. Smith (eds), *The Capitalist City. Global Restructuring and Community Politics*, Oxford: Basil Blackwell.
Fenner, F. and White, D.A. (1976), *Medical Virology*, London: Academic Press.
Figlio, K. (1979), 'Sinister medicine? A critique of left approaches to medicine', *Radical Science Journal*, 9, 14–69.
Fisher, R. (1994), *Let the People Decide. Neighbourhood Organizing in America*, New York: Twayne Publishers.
Forastieri, F., Corbo, G.M., Michelozzi, P., Pistelli, R., Agabiti, N., Brancato, G., Ciappi, G. and Perucci, C.A. (1992), 'Effects of environment and passive smoking on the respiratory health of children', *International Journal of Epidemiology*, **21**(1), 66–73.
Forrester, J.W. (1970), *World Dynamics*, Cambridge, MA: Wright-Allen Press.
Forsberg, B., Pekkanen, J., Clenchass, J. and Martensson, M.B. (1997), 'Childhood asthma in four regions in Scandinavia: Risk factors and avoidance effects', *International Journal of Epidemiology*, 26, 610–19.
Foster, J. (1992), 'The absolute general law of environmental degradation under capitalism', *Capitalism, Nature, Socialism*, **3** (3), 77–82.
Fox, A.J., Goldblatt, P.O. and Jones, D.R. (1985), 'Social class mortality differentials: artefact, selection or life circumstances', *Journal of Epidemiology and Community Health*, 39, 1–8.
Friends of the Earth (1991), *Transport and Climate Change: Cutting Carbon Dioxide Emissions from Cars*, Report for Friends of the Earth by Claire Holman, London: Friends of the Earth.
Friends of the Earth (1994), *Vehicles Emissions and Health*, London: Friends of the Earth.
Friends of the Earth (1995), *Prescription for Change. Health and the Environment*, London: Friends of the Earth.
Gandy, M. (1997), 'Postmodernism and environmentalism: complementary or contradictory discourse', in Michael Redclift and Graham Woodgate (eds), *The International Handbook of Environmental Sociology*, Cheltenham, UK and Brookfield, US: Edward Elgar.
Gans, H.J. (1982), 'The participant observer as a human being: observations on the personal aspects of fieldwork', in M. Bulmer (ed.), *Field Research: A Sourcebook and Field Manual*, London: Allen and Unwin.
Gardner, M.J. (1973), 'Using the environment to explain and predict mortality', *Journal of the Royal Statistical Society*, **136** (3), 421–40.
Gardner, M.J. (1989), 'Review of reported increases of childhood cancer rates in the vicinity of nuclear installations in the UK', *Journal of the Royal Statistical Society*, **152**, (3), 307–25.
Gardner, M.J. (1991), 'Childhood cancer and nuclear installations', *Public Health*, 105, 277–85.
Gardner, M.J., Snee, M.P., Hall, A.J., Powell, C.A., Downes, S. and Terrel, J.D. (1990), 'Results of case-control study of leukaemia and lymphoma among

young people near Sellafield nuclear plant in West Cumbria', *British Medical Journal*, February, 300, 423–34.

Gene McMullen, Gene Director (Febuary 1992), City of Houston Air Quality Control Bureau personal communication.

Ghazi, P. (1992), 'Why breathing in Britain can be damaging to your health', *The Observer on Sunday* 26 July, p. 4.

Giddens, A. (1990), *The Consequences of Modernity*, Cambridge: Polity.

Girt, J.L. (1972), 'Simple Chronic Bronchitis and Urban Ecological Structure', in N.D. McGlashan (ed.), *Medical Geography. Techniques and Field Studies*, London: Methuen.

Golding, J. (1986), 'Child health and the environment', *British Medical Bulletin*, **42** (2), 204–11.

Goldman, B.A. (1994), *Not Just Prosperity: Achieving Sustainability with Environmental Justice*, Washington, DC: National Wildlife Federation.

Goldman, B.A. (1996), 'A critical review of the methodology of environmental racism research', *Antipode*, **28** (2), 122–41.

Goren, A.I., Hellman, S., Brenner, S., Egoz, N. and Rishpon, S. (1990), 'Prevalence of respiratory conditions among school children exposed to different levels of air pollutants in the Haifa Bay area, Israel', *Environmental Health Perspectives*, 89, 225–31.

Gorz, A. (1983), *Ecology as Politics*, London: Pluto Press.

Gottdiener, M. (1985), *The Social Production of Urban Space*, Austin, TX: University of Texas Press.

Gould, K.A., Schnaiberg, A. and Weinberg, A. (1996), *Local Environmental Struggles. Citizen Activism in the Treadmill of Production*, New York: Cambridge University Press.

Gouldson, A. and Murphy, J. (1997), 'Ecological modernisation: restructuring industrial economies', in Michael Jacobs (ed.), *Greening the Millennium? The New Politics of the Environment*, Oxford: Blackwell.

Graham, N.M. (1990), 'The epidemiology of acute respiratory infections in children and adults: a global perspective', *Epidemiologic Reviews*, 12, 149–78.

Grandolfo, J. (1989), 'Blast showers neighbors with debris', *Houston Post*, 24 October, p. 1.

Greater Houston Partnership (1990a), *Here is Houston. A Newcomer's Guide. 1990*, Houston: Greater Houston Partnership: Chamber of Commerce Division, Economic Development Division and World Trade Division.

Greater Houston Partnership (1990b), *Houston Facts 1990*, Houston: Greater Houston Partnership: Chamber of Commerce Division, Economic Development Division and World Trade Division.

Greater Houston Partnership (1991), *The Houston Sourcebook. Economic and Demographic Data*, Vol. 1, Houston: Greater Houston Partnership: Chamber of Commerce Division, Economic Development Division and World Trade Division.

Greater Houston Partnership (1992), *Houston Economic Overview*, Houston: Greater Houston Partnership: Chamber of Commerce Division, Economic Development Division and World Trade Division.

Greater Houston Partnership (1995/96), *Medical Houston. Guide to the Medical Industry*, Vol. II, Houston: Greater Houston Partnership: Chamber of Commerce Division, Economic Development Division, and World Trade Division, Houston: MARCOA.

Gregory, D. (1986), 'Realism', in R.J. Johnston (ed.), *The Dictionary of Human Geography*, Oxford: Basil Blackwell.

Grubb, M. (1991), *Energy Policies and the Greenhouse Effect*, Vol. 1, Aldershot, UK: RIIA/Dartmouth Publishing.

Haan, M., Kaplan, G.A. and Camacho, T. (1987), 'Poverty and health: prospective evidence from the Alameda County Study', *American Journal of Epidemiology*, **125** (6), 959–98.

Hall, C. (1994), "Government seeks "asthma epidemic" data', *The Independent*, 18 July, p. 3.

Hardikar, Sulabaha (1992), City of Houston Health and Human Services Department, personal communication.

Hardin, G. (1977), 'The tragedy of the commons', in G. Hardin and J. Baden (eds), *Managing the Common*, San Francisco, CA: W.H. Freeman, February.

Hardoy, J. and Satterthwaite, E. (1987), *Las Ciudades del Tercer Mundo y el Medio Ambiente de la Pobreza*, Buenos Aires Grupo Editor Latinoamericano.

Harlap, S., Stenhouse, N.N. and Davies, A.M. (1973), 'A multiple regression analysis of admissions of infants to hospital: a report from the Jerusalem perinatal study', *British Journal of Preventive and Social Medicine*, 27,182–7.

Harré, R. (1970), *The Principles of Scientific Thinking*, London: Macmillan.

Hart, J.T. (1975), 'The inverse care law', in C. Cox and A. Mead (eds), *A Sociology of Medical Practice*, London: Collier-Macmillan.

Hart, N. (1986), 'Inequalities in health: The individual versus the environment', *Journal of the Royal Statistical Society*, A, 149, Part 3, 228–46.

Harvey, D. (1974), 'Population, resources, and the ideology of science', *Economic Geography*, 50, 256–77.

Hatch, M. (1992), 'Childhood leukaemia around nuclear facilities: a commentary', *The Science of the Total Environment*, 127, 37–42.

Hawley, A. (1981), *Urban Society: An Ecological Approach*, 2nd edn. New York: John Wiley.

Healey, J.H. (1990), *Statistics: A Tool for Social Research*, 2nd edn. Belmont CA: Wadsworth Publishing.

Hecht, S. and Cockburn, A. (1989), *The Fate of the Forest: Developers, Destroyers and Defenders of the Amazon*, London: Verso.

Heiman, M.K. (1996), 'Race, waste, and class: New perspectives on environmental justice', *Antipode*, **28** (2), 111–21.

Henry, R.L., Abramson, R., Adler, J.A., Wlodacyzk, J. and Hensley, M.J. (1991), 'Asthma in the vicinity of power stations: A prevalence study', *Paediatrics*, 11, 127–33.

Herbert, D.T. and Thomas, C.J. (1988), *Urban Geography: A first Approach*, 2nd edn. London: David Fulton Publishers.

Hicks, N., Moss, J. and Turner, R. (1989), 'Child poverty and children's health', in D. Edgar, D. Keane and P. McDonald (eds), *Child Poverty*, Melbourne: Allen and Unwin and Australian Institute of Family Studies.

Hill, R.C. (1977), 'Capital accumulation and urbanization in the United States', *Comparative Urban Research*, 4, 39–60.
Holgate, S.T., Djukanovic, R., Howarth, P.H., Montefort, S. and Roche, W. (1993), 'The T-cell and the airway's fibrotic response in asthma', *Chest*, 103, 125S–129S.
Holguin, A.H., Buffler, P.A., Contant, C.F., Stock, T.H. Kotchmar, D., Hsi, B.P., Jenkins, D.E., Gehan, B.M., Noel, L.M. and Mei, M. (1984), 'The effects of ozone on asthmatics in the Houston area', in Si Duk Lee (ed.), *Evaluation of the Scientific Basis for Ozone/Oxidants Standards*, Houston, TX: The Association Dedicated to Air Pollution Control and Hazardous Waste Management (APCA).
Holland, W.W., Bailey, P. and Bland, J.M. (1978), 'Long-term consequences of respiratory disease in infancy', *Journal of Epidemiology and Community Health*, 32, 256–9.
Hoppenbrouwers, T. (1990), 'Airways and air pollution in childhood: state of the art', *Lung*, Supplement: 335–46.
Houston Chronicle (1990a), 'Mexican-American infant mortality rate is unexpectedly low', 21 August, p.6.
Houston Chronicle (1990b), 'Richest, poorest areas listed', 4 May, p. 4.
Houston Post (1989), 'Reducing US infant mortality rates', 28 October, p. A35.
Houston Post (1990a), 'Airing the differences', 18 January, p. A30.
Houston Post (1990b), 'City's Fifth Ward like Third World', 30 April, p.A17.
Howarth, R. and Norgaard, R. (1992), 'Environmental valuation under sustainable development', *American Economic Review*, May, pp. 473–7.
Howe, G.M. (1963), *Atlas of Disease Mortality in the United Kingdom*, London: Nelson.
Howe, G.M. (1972), *Man, Environment and Disease in Britain*, Newton Abbot: David and Charles.
Howe, G.M. (1975), 'The geography of disease', in C.O. Carter and J. Peel, *Equalities and Inequalities in Health*, London: Academic Press.
Hsieh, K.H. and Shen, J.J. (1988), 'Prevalence of childhood asthma in Taipei, Taiwan and other Asian Pacific countries', *Journal of Asthma*, 25, 73–82.
Illich, I. (1975), *Medical Nemesis: The Expropriation of Health*, New York: Pantheon.
ILO, International Labour Office (1927), *White Lead. Data Collected by ILO in regard to the Use of White Land in the Painting Industry*, Geneva: ILO.
Ives, J.H. (ed.) (1985), *The Export of Hazard. Transnational Corporations and the Environmental Control Issue*, Boston, London and Henley: Routledge and Kegan Paul.
Jaakkola, J.J.K., Paunio, M. Virtanen, M. and Heinonen, O.P. (1990), 'Low-level air pollution and upper respiratory infections in children', *American Journal of Public Health*, **81** (8), 1060–63.
Jackle, J.A. (1994), 'Landscapes redesigned for the automobile', in M.P. Conzen (ed.), *The Making of the American Landscape*, New York and London: Routledge.
Jacobs, M. (1994), 'The limits to neoclassicism: towards an institutional environmental economics', in M. Redclift and T. Benton (eds), *Social Theory and the Global Environment*, London and New York: Routledge.

Jacobs, M., (1997), 'Introduction: The new politics of the environment', in Michael Jacobs (ed.), *Greening the Millennium? The New Politics of the Environment*, Oxford: Blackwell.

Janicke, M. (1985), *Preventive Environmental Policy as Ecological Modernisation and Structural Policy*, Berlin Science Centre: Berlin.

Janicke, M. and Weidmer, H. (1995), 'Successful environmental policy: an introduction', in M. Janicke and H. Weidmer (eds), *Successful Environmental Policy: A Critical Evaluation of 24 Cases*, Berlin: Sigma, pp. 10–26.

Johnston, R.J. (1994), *Philosophy and Human Geography. An Introduction to Contemporary Approaches*, 2nd edn. London and New York: Edward Arnold.

Jolly, D.L. (1990), *The Impact of Adversity on Child Health – Health and Disadvantage*, Victoria: The Australian College of Paediatrics.

Jones, K. and Moon, G. (1987), *Health, Disease and Society. An Introduction to Medical Geography*, London: Routledge and Kegan Paul.

Kapp, K. (1950), *The Social Costs of Private Enterprise*, New York: Schocken.

Kasarda, J.D. (1980), 'The implications of contemporary redistribution trends for national urban policy', *Social Science Quaterly*, 61, 373–400.

Katsouyani, K., Pantazopoulou, A., Touloumi, G. and Trichopoulos, D. (1993), 'Short-term effects of air pollution on mortality and hospital emergency admissions', in S. Médina and P. Quénel, *Air Pollution and Health in Large Metropolises. Technical Report*, Paris: Observatoire régional de santé d'Ile-de-France.

Keiding, L.M., Rindel, A.K. and Kronborg, D. (1995), 'Respiratory Illness in Children and Air Pollution in Copenhagen', *Archive of Environmental Health*, 50, 200–206.

Keil, R. (1995), 'The environmental problematic in world cities', in P.L. Knox and P.J. Taylor (eds), *World Cities in a World-System*, Cambridge: Cambridge University Press.

King, A.D. (1984), *The Bungalow: The Production of Global Culture*, London: Routledge and Kegan Paul.

King, A.D. (1990), *Global Cities. Post-Imperialism and the Internationalization of London*, London: Routledge.

King, A.D. (ed.) (1991), *Culture, Globalisation and the World-System*, London: Macmillan.

Kinney, P.L., Thurston, G.D. and Raizenne, M. (1996), 'The Effects of Ambient Ozone on Lung Function in Children: A Reanalysis of Six Summer Camp Studies', **104**, (2), 170–74.

Kinney, P.L., Ware, J.H., Spengler, J.D., Dockery, D.W., Speizer, F.E. and Ferris, R. G. Jr (1989), 'Short-term pulmonary function change in association with ozone levels', *American Review of Respiratory Disease*, 139, 56–61.

Klineberg, S.L. (1990a), *The Texas Environmental Survey – 1990*, Houston, TX: Rice University.

Klineberg, S.L. (1990b), 'Beyond the o. zone. Shaping up Houston', *The Architecture and Design Review of Houston*, 27: Fall.

Klineberg, S.L. (1991), *The Houston Area Survey–1982–1991*, Houston, TX: Rice University.

Klineberg, S.L. (1992), Department of Sociology, Rice University, Houston; author of the first environmental survey in Houston, personal communication.

Klineberg, S.L. (1999), 'Perspective on a city in transition', *Abode*, April 1999, p. 4

Knight, P. (1991), 'Measles epidemics, vaccine shortfall stirs' up controversy', *American Society of Medicine News*, 57, 11.

Krause, K. (1990), 'Texans say "Yes" to the environment', *Rice Reader*, 22 October.

Kripke, M.L. (1988), "Impact of ozone depletion on skin cancers", *Journal of Dermatological Surgery Oncology*, **14** (8), 853–7.

Kripke, M.L. (1989), 'Potential carcinogenic impacts of stratospheric ozone depletion', *Journal of Environmental Science and Health*, C7 (1), 53–74.

Kuhn, C.E. and Sennhauser, F.H. (1995), 'Individual Assessment of Long-Term Exposure to Ambient NO_2 and its Relation to Respiratory Symptoms in Swiss School-Children', *European Respiratory Journal*, 8, p. 286.

Laitinen, L.A., Laitinen, A. and Haahtela, T. (1993), 'Airway mucosal inflammation even in patients with newly diagnosed asthma', *American Review of Respiratory Disease*, 147, 697–704.

Laor, A., Cohen, L. and Danon, Y.L. (1993), 'Effects of time, sex, ethnic origin and area of residence on prevalence of asthma in Israeli adolescents', *British Medical Journal*, 305, 1326–9.

Lawrance, J. (1993), 'Polluted city air claims more lives', *The Times*, 9 December.

Lean, G. (1993a), 'Gasping for breath', *The Independent*, 10 October, p. 19.

Lean, G. (1993b), 'Whitehall rigs car fumes data', *The Independent*, 10 October, p. 1.

Lean, G. (1993c), 'Plot to go soft on car fumes', *The Independent*, 17 October p. 8.

Lean, G. (1993d), 'Air pollution tests increase', *The Independent*, 7 November, p. 6.

Lean, G. (1994a), 'Summer smog hovers over England', *The Independent*, 10 July, p. 1.

Lean, G. (1994b), 'Heat but no light on lethal sunshine', *The Independent*, 17 July, p. 5.

Lean, G. (1994c), 'The day Britain choked', *The Independent*, 17 July, p. 1.

Lean, G., in association with Friends of the Earth (1995), 'Where did all the fresh air go? Air pollution is one of the fastest-growing health hazards in modern Britain', *The Independent*, 5 March, 4–9.

Leeder, S.R., Corkhill, R., Irwig, L.M., Holland, W.W. and Colley, J.R.T. (1976), 'Influence of family factors on the incidence of lower respiratory tract illness during the first year of life, *British Journal of Preventive and Social Medicine*, 30, 203–12.

Leroy, P. (1995), *Environmental Science as a Vacation*, Katholieke Universiteit Netherkands: Nijmegen.

Lewis, P.R., Hensley, M.J., Wlodarczyk, J., Toneguzzi, R.C., Westley Wise, V.J., Dunn, T. and Calvert, D. (1998), 'Outdoor air pollution and children's

respiratory symptoms in the steel cities of New South Wales', *Medical Journal of Australia*, **169** (9), 459–63.
Lewontin, R.C. and Levins, R. (1996), 'False dischotomies', *Capitalism, Nature, Socialism*, **7** (3), 27–30
Lewontin, R.C. and Levins, R. (1997), 'The biological and the social', *Capitalism, Nature, Socialism*, **8** (3), 89–92.
Liebrum, J. (1990), 'Breathing can be hazardous to your health', *Houston Chronicle*, 31 October, p. 1A.
Lin-Fu, J. (1979), 'Children and lead', *New England Journal of Medicine*, 300, 731–2.
Lipietz, A. (1996), 'Geography, ecology, democracy', *Antipode*, **28** (3), 219–28.
Lonsdale, S. and Lockhart K. (1991), 'Pollution alert as freeze traps poisonous fog in towns', *The Observer*, 15 December, p. 2.
Lueunberg, P., Ackermann Liebrich, U., Kunzli, N., Schindler, C. and Perruchoud, A.P. (2000), 'Sapaldia: past, present, future', *Schweizerische Medizinische Wochenschrift*, **130**, (8) 291–7.
Lunn, J.E., Knowelden, J. and Roe, J.W. (1970), 'Patterns of respiratory illness in Sheffield infant school children', *British Journal of Preventive Social Medicine*, 24, 223–8.
MacDonald, E. (1976), 'Demographic variation in relation to industrial and environmental influence', *Environmental Health Perspectives*, 17, 153–66.
Macintyre, S., Maciver, S. and Sooman, A. (1993), 'Area, class and health: Should we be focusing on places or people?', *Journal of Social Policy*, **22** (2), 213–34.
Mahaffey, K.R., Annest, J.L. and Roberts, J. (1982), 'National estimates of blood lead levels: United States 1976–1980: Associated with selected demographic and socioeconomic factors', *New England Journal of Medicine*, 307, 573–9.
Malecki, Edward, J. (1981), 'Recent trends in location of industrial research and development', in John Rees, Geoffrey Hewings and Howard Stafford (eds), *Industrial Location and Regional Systems*, Brooklyn: J. F. Bergin, pp. 223–34.
Mallol, J. and Nogues, M.R. (1991), 'Air pollution and urinary thio-ether excretion in children of Barcelona', *Journal of Toxicology and Environmental Health*, 33, 189–95.
Mandel, E. (1978), *Late Capitalism*, London: Verso.
Martinez, F.D., Antognoni G. and Macri, F. (1988), 'Parental smoking enhances bronchial responsiveness in nine-year old children', *American Review of Respiratory Diseases*, 138, 518–23.
Martinez, F.D., Cline, M. and Burrows, B. (1992), 'Increased incidence of asthma in children of smoking mothers', *Paediatrics*, 89, 21–6.
Massey, D. (1987), *Spatial Divisions of Labour. Social Structures and the Geography of Production*, 2nd edn. London: Macmillan Educational.
Massey, D. (1995), 'Spatial Division of Labour Revisited', Lecture presented at the London School of Economics and Political Science, 16 March.
Massey, D. (1996), *Space, Place and Gender*, 2nd edn. Cambridge: Polity Press.
Massey, D. (1996) 2nd edition, *Space, Place and Gender*, Polity Press: Cambridge, UK .

Massey, D. and Meegan, R. (eds) (1985), *Politics and Methods. Contrasting Studies in Industrial Geography*, London and New York: Methuen.
McGlashan, N.D. (1972), *Medical Geography. Techniques and Field Studies*, London: Methuen.
McMullen, G. (1989), *Handy Things to Know about Houston's Air Quality*, Bureau of Air Quality Control, City of Houston, Department of Health and Human Services.
McMullen, G., Director (1992), City of Houston Air Quality Control Bureau, personal communication.
Meadows, D.H., Meadows, D.L., Randers, J. and Behrens, W.W. (1972), *The Limits to Growth: A Report for the Club of Rome's Project on the Predicament of Mankind*, New York: Universe Books.
Medical Research Council and Institute of Environmental Health (1994), *IEH Report on Air Pollution and Respiratory Disease: UK Research Priorities*, Report 2, Leicester, UK: Institute for Environment and Health.
Médina, S. and Quénel, P. (1993), *Air Pollution and Health in Large Metropolises. Technical Report*, Paris: Observatoire régional de santé d'Ile-de-France.
Mega, V. (1996), 'Our city, our future: towards sustainable development in European cities', *Environment and Urbanization*, **8** (1), 133–54.
Meyer, D.R. (1994), 'The new industrial order', in M.P. Cozen (ed.), *The Making of the American Landscape*, 2nd edn. New York and London: Routledge.
Middleton, N. (1999), *The Global Casino. An Introduction to Environmental Issues*, 2nd edn. London and New York: Arnorld.
Miller, C.A., Coulter, E.J., Schorr, L.B., Fine, A. and Adams-Taylor, S. (1985), 'The world economic crisis and the children: United States case study', *International Journal of Health Services*, **15** (1), 95–134.
Mol, A.P.J. (1997), 'Ecological modernization: industrial transformations and environmental reform', in Michael Redclift and Graham Woodgate (eds) *The Environmental Handbook of Environmental Sociology*, Cheltenham, UK and Brookefield, US: Edward Elgar.
Mol, A.P.J. and Sonnenfeld D.A. (2000), 'Ecological modernization around the world: an introduction', *Environmental Politics*, **9** (1), 3–14.
Mol, A.P.J. and Spaarsgaren, G. (2000), 'Ecological modernization theory in debate: a review', *Environmental Politics*, **9** (1), 17–49.
Montalvo, A. (1990), Responsible for Casa María Volunteers' Clinic, Southwest Houston, personal communication.
Morris, J. and Dawson, B. (1990), 'Nobody's neutral about toxic waste incinerators', *Houston Chronicle*, 22 October, p. 3.
Morris, J.N. (1975), *Uses of Epidemiology*, London: Livingston.
Murie, A. (1983), *Housing Inequality and Deprivation*, London: Heinemann Educational Books.
National Asthma Campaign (1994), *Factsheet*, London: National Asthma Campaign.
National Society for Clean Air and Environmental Protection (1995), *Air Pollution and Human Health*, London: NSCA.

Navarro, V. (1976), *Medicine under Capitalism,* New York: Proditst.

Navarro, V. (1980), 'Work, ideology and science: The case of medicine', *Social Science and Medicine,* 14c, 191–205.

Navarro, V. (1986), *Crisis, Health, and Medicine. A Social Critique,* New York and London: Tavistock Publications.

Needleman, H.L., Gunnoe, C., Leviton, A., Reed, R., Peresie, H., Maher, C. and Barret, P. (1979), 'Deficits in psychologic and classroom performance of children with elevated dentine lead levels', *New England Journal of Medicine,* 300, 689–95.

Newby, H. (1991), 'One world, two cultures: sociology and the environment', *Network,* British Sociological Association Bulletin, 50, 1–8.

Newkirk, J. (1991), 'Smog still a problem in Houston, but not as much', *Houston Post,* 22 April, p. F-4.

Nijkamp, P. and Perrels, A. (1994), *Sustainable Cities in Europe,* London: Earthscan.

Norgaard, R. (1997), 'A coevolutionary environmental sociology', in Michael Redclift and Graham Woodgate (eds) *The International Handbook of Environmental Sociology,* Cheltenham, UK and Brookfield, US: Edward Elgar.

Norgaard, R.B. (1985), 'Environmental economics: an evolutionary critique and a plea for pluralism', *Journal of Environmental Economics and Management,* 12, 382–94.

Norgaard, R.B. (1994), *Development Betrayed. The End of Progress and a Coevolutionary Revisioning of the Future,* London and New York: Routledge.

Norusis, M.J. (1988), *Introductory Statistics Guide. For SPSS-X Release 3,* Chicago, IIII: SPSS Inc.

Nyhan, W.L. (1985), 'Lead intoxication in children', *The Western Journal of Medicine,* September, 143, 3, 357–64.

Oakley, A. (1992), *Social Support and Motherhood. The Natural History of a Research Project,* Oxford, UK and Cambridge, MA.: Basil Blackwell.

O'Connor, J. (1988), 'Capitalism, nature, socialism: a theoretical introduction', *Capitalism, Nature, Socialism,* 1 (1), 16–17.

O'Connor, J. (1994), 'Is sustainable capitalism possible?', in Martin O'Connor (ed.), *Is Capitalism Sustainable? Political Economy and the Politics of Ecology,* New York and London: The Guilford Press.

O'Connor, J. (1998), *Natural Causes. Essays in Ecological Marxism,* New York and London: The Guilford Press.

O'Connor, M. (1994), 'On the misadventures of capitalist nature', in Martin O'Connor (Ed.) *Is Capitalism Sustainable? Political Economy and the Politics of Ecology,* New York and London: The Guilford Press.

OECD (1997), *Economic Globalisation and the Environment,* Paris, Organization for Economic Co-operation and Development.

Office of Population Censuses and Surveys (1973), *The General Household Survey,* Central Statistical Office, London: HMSO.

Ong, S.G., Liu, J., Wong, C.M., Lam, T.H., Tam, A.Y., Daniel, L. and Hedley, A.J. (1991), 'Studies on the respiratory health of primary school children in urban communities of Hong Kong', *Science of the Total Environment,* 106 (1–2), 121–135.

O'Riordan, T. and Turner, R.K. (1983), *An Annotated Reader in Environmental Planning and Management*, Oxford: Pergamon Press.

O'Riordan, T. and Voisey, H. (1997), 'The political economy of sustainable development', *Environmental Politics*, **6** (1) 1–23.

Ostro, B.D., Kuosett, M.J., and Mann, J.K. (1995), 'Air pollution and asthma exacerbations among African-American children in the Los Angeles area', *Inhalation Toxicology*, 7, 711–22.

Ostro, B.D., Lipsett, M.J. and Das, R. (1996), 'Particulate Matter and Asthma: A quantitative assessment of the current evidence' in *Proceedings of the Second Colloquium on Particulate Air Pollution and Human Health*, 1–3 May, Utah, 4–359–4–377.

Ostrol, B.D. (1990), 'Association between morbidity and alternative measures of particulate matter', *Risk Analysis*, 10, 421–7.

Palmer, R. and Ballantyne, A. (1991), 'Health warning as smog covers London', *The Sunday Times*, 15 December, p. 1.

Parliamentary Office of Science and Technology (1994), *Breathing in Our Cities. Urban Air Pollution and Respiratory Health*, London: POST, House of Commons.

Pearce, D. (ed.) (1991), *Blueprint 2: Greening the World Economy*, London: Earthscan.

Pearce, D.W. (ed) (1991), *Blueprint 2. Greening the World Economy*, Earthscan: London.

Pearce, D., Markandya, A. and Barbier, E.B. (1989), *Blueprint for a Green Economy*, London: Earthscan.

Pearce, F. (1992), 'Back to the days of deadly smogs', *New Scientist*, 5 December, pp. 25–8.

Pearson, A. (1990), 'Ranks of lawyers sort Phillips suits', *Houston Chronicle*, 4 November, p. 5.

Peet, R. and Thrift, N. (eds.), (1989), *New Models in Geography: The Political-Economy Perspective*, London and Winchester, MA.: Unwin Hyman.

Pepper, D. (1999), 'Ecological modernisation or the "ideal model" of sustainable development? Questions prompted at Europe's periphery', *Environmental Politics*, **8** (4), 2–34.

Perry, D.C. and Watkins, A.J. (1977), 'Regional change and the impact of uneven urban development', in D.C. Perry and A.J. Watkins (eds), *The Rise of the Sunbelt Cites*, Beverley Hills, CA: Sage.

Peters, A., Tuch, T., Brand, P., Heyder, J. and Wichmann, H.E. (1996), 'Size distribution of ambient particles and its relevance to human health', *Proceeding of the Second Colloqium on Particulate Air Pollution and Human Health*, May 1–3 1996 Park City,Utah, Jeffrey Lee (University of Utah, Salt Lake City,) and Robert Phalen (University of California) (eds), 1996, 4–406 - 4–412.

Phillimore, P. (1991), 'How do places shape health? Thinking locality and lifestyle and North-East England', Presentation at BSA Conference, 25–28 March.

Pierce, J.P., Fiore, M.C., Novotny, T.E., Hatziandreu, E.J. and Davis, R.M. (1989), 'Trends in cigarette smoking in the United States', *JAMA*, 261, 56–60.

Platt, J. (1986), 'Functionalism and the survey: the relation of theory and method', *Sociological Review*, **34**, 3, August, 501–36.

Pönkä, A. (1991), 'Asthma and low level air pollution in Helsinki', *Archives of Environmental Health*, **46** (5), pp. 263–70.

Pope, C.A. (1989), 'Respiratory disease associated with community air pollution and a steel mill, Utah Valley', *American Journal of Community Health*, **79** (5), 623–628.

Pope, C.A. III, Dockery, D.W., Spengler, J.D. and Raizenne, M.E. (1991), 'Respiratory health and PM-10 pollution: a daily time series analysis', *American Review of Respiratory Disease*, 144, 668–74.

Porritt, J. (1984), *Seeing Green: the Politics of Ecology Explained*, Oxford: Basil Blackwell.

Port of Houston Authority (1991a), *1990 Annual Report*, Houston, TX: PHA.

Port of Houston Authority (1991b), *The Houston Ship Channel. Economic Lifeline To The World*, Houston, TX: PHA.

Port of Houston Authority (1991c), *Welcome to the Port of Houston*, Houston, TX: PHA.

Port of Houston Authority (1995), *1994 Facts about the Port of Houston*, Houston, TX: PHA.

Pratt, A.C. (1991), 'Reflections on critical realism and geography', Review essay, *Antipode*, **23** (2), 248–55.

Pratt, A.C. (1994), *Uneven Reproduction. Industry, Space and Society*, London and New York: Pergamon Press.

Pratt, A.C. (1995), 'Putting critical realism to work: the practical implications for geographical research', *Progress in Human Geography*, **19** (1), 61–74.

Pratt, G. (1989), 'Reproduction, class, and the spatial structure of the city', in R. Peet and N. Thrift (eds), *New Models in Geography: the Political-Economy Perspective*, London and Winchester, MA: Unwin Hyman.

Pratt, J.A. (1980), *The Growth of a Refining Region*, Greenwich, CT: Jai Press.

Prestt, I. (1970), 'The effect of DDT on bird populations', in A. Warren and F. Goldsmith (eds) (1983), *Conservation in Perspective*, London: John Wiley.

Price, J. (1987), *Texas Procedure for Assessing Air Toxics*, Symposium presentation, TACB, Houston, 22 January.

Pulido, L. (1996), 'Volunteers, NIMBYs, and environmental justice: Dilemmas of democratic practice', *Antipode*, **28** (2), 142–59.

Pyle, G.F. (1976), 'Introduction: foundations to medical geography', *Economic Geography*, **52** (2), 95–102.

Pyle, G.F. (1979), *Applied Medical Geography*, Washington, DC: V.H. Winstons and Sons.

Quénel, P. and Médina, S. (1993), 'Short term effects of air pollution on health services', in S. Médina and P. Quénel, *Air Pollution and Health in Large Metropolises. Technical Report*, Paris: Observatoire régional de santé d'Ile-de-France.

Radford, E.P. (1976), 'Cancer mortality in the steel industry', *Annals of the New York Academy of Sciences*, 271, 228–38.

Ransom, M.R. and Pope III, C.A. (1992), 'Elementary school absences and PM-10 pollution in Utah Valley', *Environmental Research* 58, 204–19.

Read, R. and Read, C. (1991), 'Breathing can be hazardous to your health', *New Scientist,* 23 February, pp. 34–7.

Redclift, M. (1984), *Development and the Environmental Crisis. Red or Green Alternative?,* London and New York: Methuen.

Redclift, M. (1987), *Sustainable Development. Exploring the Contradictions,* London and New York: Methuen.

Redclift, M. (1992), 'Sustainable development and global environmental change. Implications of a changing agenda', *Global Environmental Change,* 2 (1), March, 32–42.

Redclift, M. (1996), *Wasted, Counting the Costs of Global Consumption,* London: Eartscan.

Redclift, M. and Woodgate, G. (1994), 'Sociology and the environment', in M. Redclift and T. Benton (eds), *Social Theory and the Global Environment,* London and New York: Routledge.

Renn, O., Webler, T. and Wiedemann, P. (eds) (1995), *Fairness and Competence in Citizen Participation. Evaluating Models for Environmental Discourse,* Dordrecht, Boston and London: Kluwer Academic Publishers.

Repetto, R. (1989), *Wasting Assets: Natural Resources in the National Income Accounts,* Washington, DC: World Resources Institute.

Rich, S. (1990), 'Best showing ever. US infant mortality rate still trails many nations', *Houston Chronicle,* 31 August, p. 14A.

Robson, C. (1993), *Real World Research: A Resource for Social Scientists and Practitioner-Researchers,* Oxford: Basil Blackwell.

Romieu, I., Cortés Lugo, M., Ruiz Velalzco, S., Sánchez, S., Memeses, F. and Hernández, M. (1993), 'Air pollution and school absenteeism among children in México city', in S. Médina and P. Quénel, *Air Pollution and Health in Large Metropolises. Technical Report,* Paris: Observatoire régional de santé d'Ile-de-France.

Romieu, I., Meneses, F., Sienra-Monge, J.J.L., Huerta, J., Ruiz Velazco, S., White, M., Etzel, R. and Hernandez Avila, M. (1993), 'Childhood asthma and ozone exposure in México city', in S. Médina and P. Quénel, *Air Pollution and Health in Large Metropolises. Technical Report,* Paris: Observatoire régional de santé d'Ile-de-France.

Romieu, I., Weitzenfield, H. and Finkelman, J. (1990), 'Urban air pollution in Latin America and the Caribbean: Health perspectives', *World Health Statistics Quarterly,* 43, 153–67.

Rose, S., Lewontin, R.C. and Kamin, L.J. (1990), *Not In Our Genes. Biology, Ideology, and Human Nature,* 4th edn. Harmondsworth: Penguin.

Rostow, W.W. (1977), 'Regional change in the fifth Kondratieff Upswing', in *In the Rise of the Sunbelt Cities,* Beverley Hills, CA: Sage.

Rutishauser, M., Ackerman, U., Braun, Ch., Gnehm, P. and Wanner, H.U. (1990), 'Significant association between oudoor NO_2 and respiratory symptoms in preschool children', *Lung,* Supplement, pp. 347–52.

Sachs, W. (1997), 'Sustainable development' in Michael Redclift and Graham Woodgate (eds), *The Environmental Handbook of Environmental Sociology*, Cheltenham, UK and Brookfield, US: Edward Elgar.

Samet, J.M., Zeger, S.L. and Berhane, K. (1995), *Particulate Air Pollution and Daily Mortality. Replication and Validation of Selected Studies*, Andover, MA: Health Effects Institute.

Saraclar, Y., Sekerel, B.E., Kalayci, O., Cetinkaya, F., Adalioglu, G., Tuncer, A. and Tezcan, S. (1998), 'Prevalence of asthma symptoms in school children in Ankara, Turkey', *Respiratory Medicine,* **29** (2) 203–7.

Sarre, P. (1987), 'Realism in practice', *Area,* **19** (1), 3–10.

Sassen, S. (1991), *The Global City. New York, London, Tokyo*, Princeton, NJ: Princeton University Press.

Savage, M. and Duncan, S. (1990), 'Space, scale and locality: a reply to Cooke and Warde', *Antipode,* 22, 1, 67–72.

Sayer, A. (1982), 'Explanation in economic geography: abstraction versus generalization', *Progress in Human Geography,* **6** (1), 69–88.

Sayer, A. (1985), 'The difference that space makes', in D. Gregory and H. Urry (eds), *Spatial Relations and Spatial Structures*, London: Macmillan.

Sayer, A. (1992), *Method in Social Science. A Realist Approach*, 2nd edn. London and New York: Routledge.

Sayer, A. and Morgan, K. (1985), 'A modern industry in a declining region: links between method, theory and policy', in D. Massey and R.A. Meegan (eds), *Politics and Method: Contrasting Studies in Industrial Geography*, London and New York: Methuen.

Scarlett, H. (1989), 'Harris Co. ranks 6th in pollution', *Houston Post,* 20 June, p. A-1.

Schnaiberg, A. and Gould, K. (1994), *Environment and Society*, New York: St Martin's Press – now Palgrave.

Schneider, D.J. and Lavenhar, M.A. (1986), 'Lead poisoning: More than a medical problem', *American Journal of Public Health*, 76, 3.

Schoon, N. (1994a), 'Smog alert issued as level of ozone hits new high', *The Independent,* 13 July, p. 2.

Schoon, N. (1994b), 'Smog, I see no smog', *The Independent,* 14 July, pp. 2–3.

Schteingart, M. (1989), 'The environmental problems associated with urban development in México City', *Environment and Urbanization,* **1** (1), April, 40–50.

Schumacher, F. (1974), *Small is Beautiful*, London: Abacus.

Schwartz, J. and Marcus, A. (1990), 'Mortality and air pollution in London: a time series analysis', *American Journal of Epidemiology,* **131** (1), 185–94.

Schwartz, J., Slater, D., Larson, T., Pierson, W. and Koenig, J. (1993), 'Particulate air pollution and hospital emergency room visits for asthma in Seattle', *American Review of Respiratory Diseases*, 147, 826–31.

Schwartz, J., Wypij, D., Dockery, D., Ware, J., Zeger, S., Spengler, J. and Ferris Jr, B. (1991), 'Daily diaries of respiratory symptoms and air pollution: methodological issues and results', *Environmetal Health Perspectives*, 90, 181–87.

Seay, G. (1990), 'Poor who use public clinics play waiting game', *Houston Chronicle,* 21 October, p. 4.

Shannon, G.W., Spurlock, C.W., Gladin, S.T. and Skinner, J.L. (1975), 'A method for evaluating the geographic accessibility of health services', *Professional Geographer*, **XXVII** (1), February, 37–41.

Shea, C.P. (1989), 'Protecting the ozone layer', in L.R. Brown, *State of the World. A Worldwatch Institute Report on Progress Toward a Sustainable Society. 1989*, New York and London: W.W. Norton.

Shelton, A.B., Rodriguez, N.P., Feagin, J.R., Bullard, R.D. and Thomas, R.D. (1989), *Houston. Growth and Decline in a Sunbelt Boomtown*, Philadelphia, PA: Temple University Press.

Shleien, B., Ruttenber, A.J. and Sage, M. (1991), 'Epidemiologic studies of cancer in populations near nuclear facilities', *Health Physics*, 61, 699–713.

Shmitzberger, R., Rhomberg, K. and Kemmler, G. (1992), 'Chronic exposure to ozone and respiratory health of children', *Lancet*, 339, 4 April, 881–2.

Simonis, U.E. (1989), 'Ecological modernization of industrial society: three strategic elements', *International Social Science Journal*, 121, 347–61.

Singh, N. (1976), *Economics and the Crisis of Ecology*, Delhi: Oxford University Press.

Sklair, L. (1994), 'Global sociology and global environmental change' in M. Redclift and T. Benton (eds), *Social Theory and the Global Environment*, London and New York: Routledge.

Smith, J., Blake, J., Grove-White, R., Kashefi, E., Madden, S. and Percy, S. (1999), 'Social learning and sustainable communities: an interim assessment of research into sustainable communities projects in the UK', *Local Environment*, 4 (2), 137–49.

Smith, N. (1984), *Uneven Development*, Oxford: Basil Blackwell.

Sobel, L.A. (ed.) (1979), *Cancer and the Environment*, Basingstoke: Macmillan – now Palgrave.

Sobral, H.R. (1989), 'Air pollution and respiratory disease in children in São Paulo, Brazil', *Social Science and Medicine*, **29** (8), 959–64.

Soja, E. (1986), 'Modernity and locality: internationalization in Greater Los Angeles', paper presented to the ESRC conference on 'Localities in an International Economy' in King, A.D. (1990), *Global Cities*. Post.

Soja, E., Morales, R. and Wolff, G. (1983), 'Urban restructuring: an analysis of social and spatial change in Los Angeles', *Economic Geography*, 59, 195–230.

Sorelle, R. (1990a), 'Too many to heal. Houston's public health care system nearing crisis point', *Houston Chronicle*, 12 September, p. 1A.

Sorelle, R. (1990b), 'US health care gets bottom summit ranking', *Houston Chronicle*, 11 July, p. 21A.

Sorelle, R. (1990c), 'Indigent mom mortality rate soars at LBJ', *Houston Chronicle*, 22 July, p. 1A.

Sorelle, R. (1990d), 'Release of task force report on infant mortality is urged', *Houston Chronicle*, 7 August, p. 2.

Spaargaren, G. and Mol, A.P.J. (1991), 'Sociology, environment and modernity', *Society and Natural Resources*, **5** (4), 323–44.

Spaargaren, G. and Vliet B. van (2000), 'Lifestyles, consumption and the environment: the ecological modernization of domestic consumption', *Environmental Politics*, **9** (1), 50–76.

Spence, J., Walton, W.S., Miller, F.J.W. and Court, S.D. (1954), *A Thousand Families in Newcastle-upon-Tyne: An Approach to the Study of Health and Illness in Children*, London: Oxford University Press.

Spivey, G.H. and Radford, E.P. (1979), 'Inner-city housing and respiratory disease in children: a pilot study', *Archive of Environmental Health*, 34, 23–30.

Spix, C. (1997), 'Ozone Is Bad for Health – But Only for Some?' *Thorax*, 52, 938–9.

SPSS Inc. (1989), *Getting Started with SPSS-X on VAX/VMS*, Chicago, IL.: SPSS Inc.

Steadman, J.R., Ross Anderson, H., Atkinson, R.W. and Maynard, R.L. (1997), 'Emergency Hospital Admissions for Respiratory Disorders Attributable to Summer Time Ozone Episodes in Great Britain', *Thorax*, 52, 958–63.

Stern, P.C., Young, O.R. and Druckman, D. (eds) (1992), *Global Environmental Change. Understanding the Human Dimensions*, Washington, DC: National Academy Press.

Stock, T.H., Gehan, B.M., Buffler, P.A., Contant, C.F., Hsi, B.P. and Morandi, M.T. (1988), 'The Houston area asthma study: a reanalysis', *Journal of the Association Dedicated to Air Pollution Control and Hazardous Waste Management*, Houston, TX, 88, 124–6.

Stratachan, D.P, Anderson, H.R., Limb, E.E., O'Neil, A. and Wells, A. (1994), 'A national survey of asthma prevalence, severity and treatment in Great Britain', *Archive of Disease in Children*, 44, 231–6.

Strauss, W. and Mainwaring, S.J. (1984), *Air Pollution*, East Kilbride, Scotland: Thomson Litho.

Stren, R., White, R. and Whitely, J. (eds) (1992), *Sustainable Cities: Urbanization and Environment in International Perspective*, Oxford: Westview Press.

Sunnucks, L. and Osorio, J. (1992), 'Environment and Development in Latin America', *Latin American Newsletters*, Confidential Report 2.

TACB (1987), *Texas Procedure for Assessing Air Toxics*, Symposium presentation by J.H. Price, J. Wiersema, T. Dydek and J.P. Henry, Houston, 22 January.

TACB (1988), *Air Quality in Texas*, Austin, TX: Texas Air Control Board.

TACB (1988), *Carbon Monoxide. Historical Summary Report 1973–1986*, Austin, TX: Texas Air Control Board.

TACB (1992a), *Excerpt from: Texas Air Quality Compared to National Levels 1990*, Austin, TX: Texas Air Control Board.

TACB (1992b), *Air Quality in the Houston Metropolitan Area*, Austin, TX: Texas Air Control Board.

TACB (1992c), *1990 Air Quality National, Texas: A Comparison*, Austin, TX: Texas Air Control Board.

TACB (1992d), *Data Summary Supplement to the Initial Analyses of Ambient Toxics and Volatile Organic Compound Data, Private Network Data*, Data Management and Analysis Division, Austin, TX: Texas Air Control Board.

TACB (1992e), *General Rules (31 TACB Chapter 101)*, Austin, TX: Texas Air Control Board.

TACB (1993a), *Air Monitoring Report 1990* Brian Lambeth, Monica Havelka, Larry Butts and Technical Operations Staff, Austin, TX: Texas Air Control Board.

TACB (1993b), *TACB Bulletin*, No. 1, January–February, Austin, TX: Texas Air Control Board.

Tager, I.B., Weiss, S.T., Munoz, A., Rosner, B. and Speizer, F.E. (1983), 'Longitudinal study of the effects of maternal smoking on pulmonary function in children', *New England Journal of Medicine* 309, 699–703.

Taylor, A.N. (1995), 'Something in the air', *MRC News* (UK Medical Research Council), Summer, pp. 22–5.

Terblanche, A.P., Opperman, L., Nel, C.M., Reinach, S.G., Tosen, G. and Cadman, A. (1992), 'Preliminary results of exposure measurements and health effects of the Vaal Triangle Air Pollution Health Study', *South African Medical Journal*, **81** (11), 550–6.

Texas Cancer Council, the Texas Department of Health and the University of Texas M.D. Anderson Cancer Center (1991), *Impact of Cancer on Texas*, Houston: The University of Texas M.D. Anderson Cancer Center.

Thomas, R.D. and Murray, R.W. (1991), *Progrowth Politics. Change and Governance in Houston*, Institute of Governmental Studies and University of California at Berkeley: IGS Press.

Thrift, N. (1988), 'The geography of international economic disorder', in D. Massey and J. Allen (eds), *Uneven Re-Development: Cities and Regions in Transition*, London: Open University and Hodder and Stoughton.

Todaro, M.P. (1994), *Economic Development*, 2nd edn. New York and London: Longman.

Townsend, P. (1979). *Poverty in the United Kingdom. A Survey of Household Resources and Standard of Living*, Great Britain: Pinguing Books.

Townsend, P. (1983), *Poverty in the United Kingdom. A Survey of Household Resources and Standards of Living*, 3rd edn. Harmondsworth, UK: Penguin Books.

Townsend, P., Phillimore, P. and Beattie, A. (1988), *Health and Deprivation: Inequality and the North*, London: Croom Helm.

Toyoma, T. (1964), 'Air pollution and its effects in Japan', *Archive of Environmental Health*, 8, 153–73.

Turner, W.C. (1964), 'Air pollution and respiratory disease', *Proceedings of the Royal Society of Medicine*, 57, p. 618.

UNEP and WHO, United Nations Environment Programme and World Health Organization (1992), *Urban Air Pollution in Megacities of the World*, Oxford: Basil Blackwell.

Urrutia-Rojas, X. (1988), 'Health Care Needs and Utilization among Hispanic Immigrants and Refugees in Southwest Houston', MSc thesis, University of Texas Health Science Center at the Houston School of Public Health.

Urry, J. (1987), 'Society, space and locality', *Environment and Planning D: Society and Space*, 5, 4, 435–44.

US EPA (1986), *Air Quality Criteria for Lead*, US Environmental Protection Agency IV: 13.35, Washington, DC: Office of Health and Environmental Assessment.

US EPA (1990a), *National Air Quality and Emissions Trends Report, 1988*, Washington, DC: Office of Air Quality Planning and Standards, US Environmental Protection Agency.

US EPA (1990b), *Cancer Risk from Outdoor Exposure to Air Toxics*, I, Washington, DC: US Environmental Protection Agency.

US EPA (1991), *National Air Quality and Emissions Trends Report, 1990*, Washington, DC: Office of Air Quality, Planning and Standards, US Environmental Protection Agency, 450/4-91-023.

US EPA (1992), *Respiratory Health Effects of Passive Smoking. Lung Cancer and Other Disorders*, US Environmental Protection Agency 600/6-90/006F, Washington, DC: Office of Health and Environmental Assessment.

US EPA (1993), *Measuring Air Quality: The Pollutant Standards Index*, US Environmental Protection Agency 451/k-94-001, Washington, DC: Office of Air Quality Planning and Standards.

US EPA (1994), *National Air Quality and Emissions Trends Report, 1993*, US Environmental Protection Agency 454/R-94-026, Washington, DC: Office of Air Quality Planning and Standards.

US EPA (1998), *National Air Quality and Emissions Trends Report, 1996*, Office of Air Quality Planning and Standards, Washington DC, EPA 454/R-97-013.

US EPA (2000), 'EPA and DOJ announce record clean air agreement with major petroleum refiners', Headquarters Press Release, 25/7/2000.

US EPA AIRS, *Aerometrics Information Retrieval System, AMP 450*, Washington, DC.

Vaus, D.A. de (1986), *Surveys in Social Research*, London: Allen and Unwin.

Vigar, G. (2000), 'Local "barries" to environmentally sustainable transport planning', *Local Environment*, 5, 1, February, 19–32.

Villalbí, J.R., Martí, J., Aulí, E., Conillera, P. and Millá, J. (1984), 'Morbididad respiratoria y contaminación atmosférica', *Medicina Clínica* (Barcelona), 82, 695–7.

Ward, P.M. (1990), *México City. The Production and Reproduction of an Urban Environment*, London: Belhaven Press.

Watts, M. (1983), 'On the poverty of theory: natural hazards research in context', in K. Hewitt (ed.), *Interpretations of Calamity from the Viewpoint of Human Ecology*, Boston: Allen and Unwin.

WCED (1987), *Our Common Future*, The Brundtland Report, Oxford: Oxford University Press.

Weale, A., O'Riordan, T. and Kramme, L. (1991), *Controlling Pollution in the Round. Change and Choice in Environmental Regulation in Britain and West Germany*, Anglo-German Foundation for the Study of Industrial Society: Staples Printers Rochester.

Weitzman, M., Gortmaker, S., Sobol, A.M. and Perrin, J.M. (1992), 'Recent trends in the prevalence and severity of childhood asthma', *Journal of the American Medical Association*, 268, 2673–7.

Weitzman, M., Gortmaker, S., Walker, D.K. and Sobol, A. (1990), 'Maternal smoking and childhoold asthma', *Paediatrics*, 85, 505–11.

West, R. (1991), *Computing for Psychologists. Statistical Analysis Using SPSS and Minitab*, London and Paris: Hardwood Academic Press.

WHO, World Health Organization (1979), *Chronic Respiratory Diseases in Children in Relation to Air Pollution, Report on a WHO Study*, Geneva: Regional Office for Europe.

Wihill, C. (1994), Poor at greatest risk from breast cancer, *The Guardian*, 29th September, p. 8.

Williamson, H.F., Andreano, R.L., Daum, A.R. and Klose, G.C. (1963), *The American Petroleum Industry. The Age of Energy 1899–1959*, Evanston, IL.: Northwestern University Press.

Williamson, J.G. (1965), 'Regional inequality and the process of national development: A description of the patterns', *Economic Development and Cultural Change*, **13** (4), 3–84.

Wilson, D. (1983), *The Lead Scandal*, London: Heinemann Educational.

World Resources Institute (1988), *World Resources 1988–1989*, International Institute for Environment and Development and United Nations Environment Programme, New York and Oxford: Oxford University Press.

World Resources Institute (1994), *World Resources 1994–1995*, International Institute for Environment and Development and United Nations Environment Programme, New York and Oxford: Oxford University Press.

Wyckoff, W.K. (1994), 'Landscapes of private power and wealth', in M.P. Cozen (ed.), *The Making of the American Landscape*, New York and London: Routledge.

Wyne, B. (1994), 'Scientific knowledge and the global environment', in M. Redclift and T. Benton (eds), *Social Theory and the Global Environment*, London and New York: Routledge.

Yang, C.Y., Chiu, J.F., Chiu, H.F. and Kao, W.Y. (1997), 'Damp housing conditions and respiratory symptoms in primary school children' *Pediatric Pulmonology*, **24** (2), 73–7.

Yardley, M. (2000), 'Sweet stench of growth', *Houston Chronicle*, 6 April.

Yarnell, J.W.G. and Leger, A.S. (1977), 'Housing conditions, respiratory illness and lung function in children in South Wales', *British Journal of Preventive Medicine*, 31, 183–8.

Yearley, S. (1991a), 'Greens and science: a doomed affair?', *New Scientist*, 13 July, pp. 37–40.

Yearley, S. (1991b), *The Green Case. A Sociology of Environmental Issues, Arguments and Politics*, London: HarperCollins.

Yearley, S. (1994), 'Social movements and environmental change', in M. Redclift and T. Benton (eds), *Social Theory and the Global Environment*, London and New York: Routledge.

Yearley, S. (1996), *Sociology, Environmentalism, Globalization. Reinventing the Globe*, London: Sage.

Index

The manufacturer's authorised representative in the EU is Springer Nature Customer Service Centre GmbH, Europaplatz 3, 69115 Heidelberg, Germany. If you have any concerns regarding our products, please contact ProductSafety@springernature.com

Printed and bound by CPI Group (UK) Ltd, Croydon, CR0 4YY
13/10/2026
02232378-0009